D0907756

Lewis Mumford and the Ecological Region:

THE POLITICS OF PLANNING

CRITICAL PERSPECTIVES
A Guilford Series

Edited by
DOUGLAS KELLNER
University of Texas, Austin

Lewis Mumford and the Ecological Region:

THE POLITICS OF PLANNING

MARK LUCCARELLI

THE GUILFORD PRESS
New York   London

© 1995 The Guilford Press
A Division of Guilford Publications, Inc.
72 Spring Street, New York, NY 10012

All rights reserved

No part of this book may be reproduced, stored in a retrieval system,
ortransmitted, in any form or by any means, electronic, mechanical,
photocopying, microfilming, recording, or otherwise, without written
permission from the Publisher.

Printed in the United States of America

This book is printed on acid-free paper.

Last digit is print number: 9 8 7 6 5 4 3 2 1

Library of Congress Cataloging-in-Publication Data
Luccarelli, Mark.
    Lewis Mumford and ecological region : the politics of planning
    / by Mark Luccarelli.
        p.   cm. — (Critical perspectives)
    Includes bibliographical references and index.
    ISBN 1-57230-001-9 (hard). – ISBN 1-57230-228-3 (pbk.)
    1. Regional planning—United States—History.   2. Regional
Planning Association of America—History.   3. Mumford, Lewis 1895–
1992.   I. Title.   II. Series: Critical perspectives (New York, N.Y.)
HT392.L92   1995
711′.4′092—dc20   95-30860
                                                                  CIP

Designed and formatted by
KP Company
Brooklyn, NY

For my parents

Carmela and Antonio Luccarelli

Figure Sources

FIGURE 1. From John W. Reps, *Town Planning in Frontier America*, Princeton University Press, 1969, page 156. Source: *Salem, Massachusetts: 1670*. Manuscript redrawing by John Gibson in 1962 of a plan of Salem, Massachusetts, in 1670, from Sidney Perley, *The History of Salem Massachusetts*. Salem, 1924, Vol. 1. Division of Rare and Manuscript Collections, Cornell University Library.

FIGURE 2. From John W. Reps, *Town Planning in Frontier America*, Princeton University Press, 1969, page 211. Source: *A Portraiture of the City of Philadelphia in the Province of Pennsylvania in America*. Plan of Philadelphia, Pennsylvania, drawn by Thomas Holme. From a restrike in John C. Lowber, *Ordinances of the City of Philadelphia, 1812*. Philadelphia, 1812. Division of Rare and Manuscript Collections, Cornell University Library.

FIGURE 3. From Benton MacKaye, *The New Exploration: A Philosophy of Regional Planning*, University of Illinois Press, 1962, page 199. Copyright © 1995 Christie MacKaye Barnes. Reprinted by permission.

FIGURE 4. From Benton MacKaye, *The New Exploration: A Philosophy of Regional Planning*, University of Illinois Press, 1962, page 172. Copyright © 1995 Christie MacKaye Barnes. Reprinted by permission.

FIGURE 5. From Benton MacKaye, *The New Exploration: A Philosophy of Regional Planning*, University of Illinois Press, 1962, page 173. Copyright © 1995 Christie MacKaye Barnes. Reprinted by permission.

FIGURE 6. From Benton MacKaye, *The New Exploration: A Philosophy of Regional Planning*, University of Illinois Press, 1962, page 185. Copyright © 1995 Christie MacKaye Barnes. Reprinted by permission.

FIGURE 7. From John I. Bright, "Unpopulating London," *Journal of the American Institute of Architects* 8 (October 1920): 354–356. Diagram prepared by the Garden Cities and Town Planning Association, London. Rare Books, Firestone Library, Princeton University.

FIGURE 8. From *Planning the Fourth Migration: The Neglected Vision of the Regional Planning Association of America*, edited by Carl Sussman, The MIT Press, 1976, page 188. Source: New York State Report of the Commission of Housing and Regional Planning (May 7, 1926).

FIGURE 9. From Mel Scott, *American City Planning Since 1890: A History Commemorating the Fiftieth Anniversary of the American Institute of Planners*, University of California Press, 1971, page 218. Copyright © 1969, by The Regents of the University of California. Reprinted by permission.

FIGURE 10. From Clarence S. Stein, *Toward New Towns for America*, 3rd ed., The MIT Press, 1966, page 25. Copyright © 1969 MIT Press, Cambridge, Massachusetts. Reprinted by permission.

FIGURE 11. From Clarence S. Stein, *Toward New Towns for America*, 3rd ed., The MIT Press, 1966, page 25. Copyright © 1969 MIT Press, Cambridge, Massachusetts. Reprinted by permission.

FIGURE 12. *From Regional Plan of New York and Its Environs. Volume II. The Building of the City*, page iii. Copyright © 1931 Committee on Regional Plan of New York and Its Environs. Reprinted by permission.

Acknowledgments

The creation of a book is a collective effort; only by writing this book did I learn the truth of this maxim.

I am grateful to the following libraries for assistance: University of Iowa, Van Pelt Library of the University of Pennsylvania, Olin Library of Cornell University, Firestone Library of Princeton University, and Alexander Library of Rutgers University. Elsie Luccarelli and Fanny and Tony Sasso permitted me to live with them while I did my initial research.

Many people read the manuscript at various stages and made helpful comments and corrections: James Harris, Steven Best, Pierre Clavel, Frank Popper, Robert Fishman, John L. Thomas, and Jonathan Feldman.

Thanks to Elliott Sclar for our conversations about cities and planning.

I thank my father, Antonio Luccarelli, artist and architect, whose

vii

interest in cities and architectural reform has been a cornerstone of my work, and my mother, Carmela D. Luccarelli, for her editorial assistance in the early stages.

Without the technical assistance of Richard Vanasse, this project would not have been possible.

I am grateful to Peter Wissoker of The Guilford Press for believing in the project and seeing it into production. Alice Colwell made many improvements to the final manuscript, and Rowena Howells of Guilford skillfully guided the project through its final stages.

Finally, I want to acknowledge the critically important contributions of three friends: Daniel Postel, with whom I had many important discussions about Mumford and who helped me to find people interested in the project; Jennifer Hohman, with whom I consulted on every chapter and who made helpful comments; and most importantly, the late Sherman Paul, for his invaluable insights and his model of committed scholarship and teaching.

Contents

Introduction

Lewis Mumford's career as a critic and social commentator spanned several decades. From the 1920s through the 1960s, his work addressed the environmental, aesthetic, and social dimensions of urban and regional planning. As a founding member of the Regional Planning Association of America (RPAA), Mumford advanced a concept of regional development that accounted for the mutual importance of the social world and the natural ecosystem. In his unique interpretation of the role of "technics" in "civilization," Mumford linked the planning of towns and cities to an empathetic understanding of the complex natural region. The natural region, with its characteristic local variations, could be saved only through a new (actually very old) kind of urban and regional planning conceived as an instrument of a civic-minded social order. Planning for ecological regionalism required cultural and political renewal.

The first part of this book traces the intellectual sources of Mumford's ecological regional vision. The idea of regionalism that Mumford counterposed to the power of the "megamachine" reflects a tradition of ecological thinking that Sherman Paul has identified as the "green tradition" in American letters.[1] This tradition, which found its most eloquent elaboration in the writings of Ralph Waldo Emerson, Henry David Thoreau and Walt Whitman, offered a critique of possessive individualism. Mumford's reading of the Transcendentalists in *The Golden Day* underscores this theme of cultural renewal through lived experience. Their exploration of experience, which bridges the subjective and natural worlds, suggests an alternative to the prevalent antisocial form of American individualism; this alternative, which Mumford advocated, links subjectivity to place. Like Charles Olson, Mumford wanted to recover the pragmatic and intuitive senses of relation to the ecology of place. His original contribution was to link this reading of the green tradition to the new technique of regional planning and in turn contribute to the rediscovery of America as a republic of regions whose geographic diversity was consonant with the American ideal of decentralized democracy.

Like his colleague Benton MacKaye, Mumford saw regionalism as "exploration," a generous "science" that made room for subjective experience. What is essential in regionalism is its vision of an organic order that enlivens culture. As such it is a response to the predominant modern Western worldview that has turned nature into empty space and promoted technological "solutions" that engulf the complexities of both the urban and natural worlds. Mumford called the outcome the "anti-city" (Henry Adams called it the "dynamo"): an increase in physical force made possible by the machine and directed against organic and human communities. Regionalism addresses this threat by recalling an organic order arising out of the experienced and observed qualities of the land. But for Mumford, the organic principle did not stand against modernity. Rather, he linked organicism to the potential of science for recovering a healthy relation to nature, as well as the American civic tradition of democratic participation.

The realization of regionalism called for regional planning. It meant using planning to adapt technology to natural patterns so as to respect limits and diversity. Then as now, regional planning must ad-

[1]Sherman Paul, *Repossessing and Renewing:Essays in the Green American Tradition* (Baton Rouge: LSU Press, 1976).

dress two phenomena: the destruction of nature and the decline of urban life. Such an integrated approach would involve

- a way to shape human life in response to the influences and critical forces of the regional ecosystems to which we must adapt.

- a movement to reorient urban life, to address the crisis of overurbanization, and to recontextualize cities in relation to nature. (Mumford cultivated the possibility of a third way: not rural or urban—at least not as these have been defined as polar opposites in the modern era—and certainly not the "middle landscape" of the pastoral suburb; rather, the city as polis in relation to the organic complexities of the regional ecosystem.)

- a means to take advantage of natural economies (rather than economies of scale), to develop sources of power that do not require fossil fuels, to discover efficiency in a workable small scale; an economic regionalism to "combat the evil of over-specialization" and engender systems of production aimed at accepting the limits of human wants and focusing on satisfying basic needs rather than perfecting a system of commodification designed for an ever expanding increase in levels and expectations of consumption.

As a founder of the Regional Planning Association of America, Mumford took public positions on the critical urban issues of his day. Outside of urban planning circles, the RPAA may be a little-known chapter in American history, but the country's failure to heed its innovative and moderate approach to regional and technological planning is an important story. It helps explain the demise of our cities as well as the depth of our environmental dilemma.

The second part of this book concerns Mumford's effort, through his association with the RPAA, to make regionalism a guiding principle of American urban development. The 1920s and 1930s were a critical time for the United States, as the ideas and perspectives of the Progressive Era came to fruition in the New Deal of the 1930s, arguably the last great attempt at political and social reform in the

United States. The RPAA, by force of its ideas alone, presented a powerful alternative both to the social and cultural pattern of suburban life and to the relation of the built environment to the natural region. My narrative is an attempt to recover the pragmatic creativity of the RPAA, which may be regarded as an example of what Van Wyck Brooks called a "usable past." It is a part of the American past that has been lost in the narrow confines of historical "development," and its value as an expression of an alternative America may therefore be easily neglected in the historical discourse.

The RPAA presented a coherent and viable alternative to megalopolis. Its plans for "regional cities" and managed decentralization took modern technologies into account and addressed the social inequities created by industrialization and capital accumulation. The group's plans put conservation of land and landscape into the context of new urban forms, reinventing the urban form to take advantage of technological innovation and to undo the damage done by industrialization. Whatever criticisms can be made of RPAA plans for the regional city—and there are several—the one indisputable achievement of the group was its recognition that we cannot have an environmentally sound policy toward land, landscapes, or ecosystems without addressing the breakdown of the human ecology, that is, the breakdown of urban forms and communities, and our failure either to restore old forms or invent new ones. The RPAA stood for inventing a new form, the garden city, which I argue was an eminently reasonable response to the problems presented by the imminent prospect of mass suburbanization. That the group ultimately failed to effect its vision represents a moral shortcoming within the country, a lost opportunity for creative reform in the United States.

Yet the RPAA's lack of success also rests with the group itself, particularly in Mumford's overoptimistic assessment of the possibilities for managing and directing technological change toward social and environmental ends. The RPAA never generated the public support necessary to realize its vision in the face of the intransigence of American elites. The result was that a decentralized regionalist alternative to the present modern political economy was never given serious public consideration.

In Part III we see that as Mumford developed his critique of power (partly in response to his disappointments with New Deal urban and economic policies), he saw the need to create those alternative institutions. He moved toward conclusions that are significant today: (1) the need to develop a public discourse and civic consciousness as a means not of dissolving but of redirecting the professionalization of knowledge; (2) the ability to find a medium to make aesthetic concerns public issues and therefore make discourse about beauty a criterion of public life; (3) the need to democratize economic and political power as an accompaniment to any attempt to reconfigure the built environment.

In the end the ideas of Lewis Mumford and the RPAA are important as instances of pragmatic American response to the conditions of modernity. I hope that my presentation here reminds us that we are not inevitably stuck in what Walter Benjamin called the "fossils" of a rigid, commodified culture; the wheels of change may turn again.[2] The fundamental supposition and value of Mumford's planning work is that if meaningful social and political change is possible, it is enabled by imaginative vision. If we are to address the twin crises of urban decay and natural destruction, we must have what Donald Worster called the "ecological imagination."[3]

[2]See Susan Buck-Morss, *The Dialectics of Seeing: Walter Benjamin and the Arcades Project* (Cambridge: MIT Press, 1990), pp. 56, 160–161.

[3]Donald Worster, *The Wealth of Nature* (Oxford: Oxford University Press, 1993), pp. 209–210.

PART I

Developing A Sense of Place

Our place is part of what we are. Yet even a "place" has a
kind of fluidity: it passes through space and time—
"ceremonial time." . . . A place will have been a grasslands,
then conifers, then beech and elm. It will have been half
riverbed, it will have been scratched and plowed by ice.
And then it will be cultivated, paved, sprayed, dammed,
graded, built up. But each is only for a while, and that will
be just another set of lines on the palimpsest. . . .

• • • •

One's sense of the scale of place expands as one learns the
region. The young hear further stories and go for explora-
tions which are also subsistence forays—firewood gathering, fishing, to fairs or to market. The outlines of the larger
region become part of their awareness. (Thoreau says in
"Walking" that an area twenty miles in diameter will be
enough to occupy a lifetime of close exploration on foot—
you will never exhaust the details.)

Gary Snyder
The Practice of the Wild

CHAPTER 1

The Old Order and the New

Lewis Mumford came of age during World War I, and his thought was occasioned by that crisis of civilization. As early as 1915 he wrote: "The methods of doing away with life have been extended by leaps and bounds. . . . From this it follows that life is becoming less precious; that disrespect for life is increasing; and that unless we look very sharply after our industrial and military systems they will between them reduce life to a low ebb."[1] The war generated two dilemmas that shaped Mumford's ideas regarding the relationship between culture and politics: one was the predicament experienced by the progressive supporters of the liberal state and the war effort; the second was the fragmentation of the prewar Left. Mumford defined himself in relation to both progressive politics and cultural radicalism. His response to the failure of both was to call for a synthesis of a politics of cultural transformation and a new science of regional geography.

The war destroyed the union of cultural and political radicalism

[1]Mumford, untitled, unpublished MS, 4 December 1915, Lewis Mumford Papers, Van Pelt Library, University of Pennsylvania.

fostered by the Young Intellectuals, or Young Americans, of the pre-war era.[2] These radicals rejected Victorian culture in favor of an alternate set of values, including spontaneity, self-expression, artistic freedom, and sexual liberation. While often identified with bohemianism, the cultural politics of the Young Intellectuals was more than a matter of style: the Young Intellectuals joined cultural radicalism with either political reformism or radicalism. As Malcolm Cowley points out in his memoir, *Exile's Return*, the "two currents"—cultural rebellion and political dissent—"were hard to distinguish."[3] They came together in journals intended to awaken the educated public. The Young Intellectuals published in the *Dial* (briefly), the *Freeman*, the *Seven Arts*, and most importantly, in the *Masses* and the *New Republic*. One intellectual axis was created by the *New Republic*, founded in 1914 by Herbert Croly: it combined the Young Intellectual agenda of "national cultural renewal" with a reformist political orientation.[4] Another axis was the *Masses*, the most radical of the Young Intellectual publications, having supported the revolutionary cause of the International Workers of the World (IWW).[5] When the United States entered the war in 1917, the Young Intellectuals associated with the *Masses* took a principled stance against it, and its editor, Max Eastman, as well as writers Randolph Bourne and Floyd Dell, were later indicted under the Sedition Act.

The Young Intellectuals' antiwar stance was their final act of defi-

[2]See John P. Diggins, *The American Left in the Twentieth Century* (New York: Harcourt, Brace, 1973); also see Henry F. May, *The End of American Innocence* (New York: Oxford University Press, 1959); and Casey Blake, *Beloved Community: The Cultural Criticism of Randolph Bourne, Van Wyck Brooks, Waldo Frank, and Lewis Mumford* (Chapel Hill: University of North Carolina Press, 1990).

[3]See Malcolm Cowley, *Exile's Return* (New York: Penguin, 1976; orig. pub. 1934), p. 66.

[4]Thomas Bender, *New York Intellect: A History of Intellectual Life in New York City, from 1750 to the Beginnings of Our Own Time* (Baltimore: Johns Hopkins University Press, 1987), p. 227, says that it "symbolized a New York–Washington reformist political axis that would be prominent again in the 1930s and 1960s."

[5]Cowley, *Exile's Return*, p. 66.

ance. The political consequences of U.S. entry into World War I effectively destroyed the bridge they had created between cultural transformation and the politics of social change. Perhaps it is true, as Thomas Bender argues, that despite their "rhetoric of justice . . . [the radicals] seemed always to . . . [have] a greater concern about repression of personal desire and creative talent in America than about class oppression."[6] In this regard it is important to consider the critical inclination of Young Intellectuals writing in the postwar period. In what was perhaps their last significant collective effort, the publication of *Civilization in the United States* (1922), writers including Van Wyck Brooks, Paul Rosenfeld, Waldo Frank, and Mumford centered their critique on what the book's editor called a "spiritually impoverished civilization."[7] The Young Intellectuals had become the "civilization group," which shared Brooks's hope that America could be reformed by cultivation of the arts and humanities.[8]

The difficulty of sustaining the Young Intellectuals' antebellum synthesis of social activism and artistic creativity should not blind us to its potential for effecting cultural transformation.[9] Nor should

[6]Bender, *New York Intellect*, p. 231. See also Christopher Lasch, *The New Radicalism* in America (New York: Knopf, 1965).

[7]Harold E. Stearns, ed., *Civilization in the United States: An Enquiry by Thirty Americans* (New York: Harcourt, Brace, 1922); Mumford's essay is entitled "The City."

[8]The importance of his participation in this project to Mumford's intellectual development is suggested by his correspondence with Brooks. Mumford's association with Brooks is described in two letters, Van Wyck Brooks to Lewis Mumford, 2 April 1920, and n.d. (1922 or 1923), Mumford Papers. Brooks, Mumford, and Rosenfeld later collaborated on the publication of *The American Caravan*, an annual anthology of works by young American writers.

[9]In *The Agony of the American Left* (New York: Knopf, 1969), Christopher Lasch suggests that the failure of the American Left has been that it has neglected to develop an oppositional liberalism that is grounded in tough-minded personal ethics and builds responsibility for society from personal responsibility. But in *Beloved Community,* Casey Blake argues that the tender-minded desire to re-create "community"—an outgrowth of *a psychology* that emphasizes the "intuitive values of the soul" may become an important basis for oppositional politics.

we overlook the constraints of historical circumstance: U.S. participation in the war brought on a repressive state apparatus and helped
foreclose radical and reformist political tendencies alike. All leftist
groups, like the American labor movement, were badly set back by
the wartime crisis, and the Young Americans were no exception.
Their attempt to draw on the American tradition of decentralization to develop an alternative, radically democratic liberalism was
stillborn.

While many leftists (including the Socialist Party of the United
States) opposed the war, the predominate tendency in progressive
circles was accommodationist, as best exemplified by John Dewey's
endorsement of American intervention. Dewey's work had been an
important influence on that of the Young Intellectuals. But in 1918
Dewey argued for political "realism": by supporting the war,
progressives would position themselves to turn the general mobilization into support for social reform. Dewey approved of the much
larger role now played by the federal government in the economy
and society; the war emergency necessitated direction of the railways, distribution of foodstuffs, and construction of housing. These
activities, he thought, might shape a shared understanding that such
wider social ends are in the public's interest and could in fact greatly
broaden the conception of public interest.[10] The effect of Dewey's
endorsement of the war was to marginalize Bourne and the other
Young Americans who opposed the war effort. In "Twilight of Idols,"
his slashing critique of Dewey's prowar stance, Bourne questioned
the idea of "instrumentalism": what good is a pragmatic politics if it
means abandonment of one's ethical principles?[11]

Less than ten years later, Mumford would repeat Bourne's charges
against Dewey,[12] but at the time Mumford's attitude toward the war—

[10]This was precisely the position of a group of socially minded architects led by Charles Harris and Frederick Ackerman, early advocates of an
extensive federal wartime housing program and leaders of the group that
would later form the core of the RPAA.

[11]Randolph Bourne, "The Twilight of the Idols," *The Seven Arts* 2, (1917),
pp. 688–702; reprinted in C. Resek, ed., *War and the Intellectuals: Essays by
Randolph S. Bourne* (New York: Harper & Row, 1964), pp. 53–64.

[12]In Mumford, *The Golden Day* (New York: Harcourt Brace, 1926).

and toward politics—was shaped by the logic of Dewey's accommodationist stance. When the war came, Mumford did not join the radical opposition: he enlisted in the Navy. This reflected a reformer's trust in the state and a belief in "the noble purposes" with which President Woodrow Wilson justified U.S. intervention. Mumford hoped that the war might establish "a new footing of peacedom."[13]

Yet Mumford was also critical of the political repression that accompanied the war effort. He noted that "the hysterical suppression of free discussion" by the Wilson administration was an "atavistic demand for complete [and false] spiritual unity."[14] The administration's heavy-handedness at home and its failure to shape a new world order at Versailles forced Mumford to conclude that the war had been simply another exercise in power politics. He remarked that the Treaty of Versailles brought the United States "one step further away from her democratic profession and anti-imperialist traditions." Thus by 1919 he had reversed his earlier endorsement of Wilsonian progressivism. Sounding more and more like Bourne, Mumford decided that war was "constitutional" to the state, which is by nature "imperialistic."[15] The war taught Mumford to be critical of the state, a position that he maintained throughout his life and that distinguished his politics from both conventional liberalism and Marxism.

Yet Mumford's criticism of the State also presented problems of consistency. On the one hand, like Bourne, he developed an ethical stance highly critical of organized power. On the other, like Dewey in 1918, he defined politics as a pragmatic instrument of social reform. This tendency toward a pragmatic politics was also apparent in his involvement with reformist planning and housing movements and in his stance in the late 1930s, when, as one of the leading "war

[13]Mumford, "The Year 1917," personal memoir, December 1917; and "This War as an Unparalleled Disaster: A Second Thought," unpublished MS, 1915, Mumford Papers.

[14]Mumford, "The Year 1917."

[15]Mumford, "Patriotism and Its Consequences," Dial, 19 April 1919, pp. 406–407; "The Old Order and the New," unsigned editorial, Dial, 26 July 1919, p. 65; "Wardom and the State," Dial, 4 October 1919, pp. 303–305.

intellectuals," he became an early proponent of American intervention against Nazi aggression.[16] Mumford's political position, therefore, reflected conflicting demands—for a politics of reform (in
response to powerful social and economic forces) and for a moral
politics of opposition (in response to the corrupting influence of raw
power). Addressing this dilemma, Mumford worked to reshape the
confines of the "political"; he knew (perhaps intuitively) that the revival of democratic politics must draw on a new relationship between art and everyday life. Thus, Mumford began by directing his
radical impulses into cultural criticism. This opening was fundamental to Mumford's work. His effort to give form to a democratic cultural vision led him over time to adapt his concept of regionalism
by encompassing both the immediacy of the local public sphere and
the pragmatic power of state planning. In effect, Mumford developed a more sophisticated politics as he became aware that the regionalist alternative he advocated required the recovery of the polis,
the public sphere, as the basis of a democratic culture.[17]

Born in New York City in 1895, Lewis Mumford was an illegiti-
mate child deserted by his father. He and his mother, Elvina, lived
with her mother and stepfather. For a few years, Elvina operated an
unsuccessful boarding house on West 65th Street. When her stepfather died in 1906, Elvina and Lewis were forced to move to a run-

[16]Mumford first advocated a tough response to Germany in "When
America Goes to War," *Modern Monthly*, June 1935, pp. 203–204. A few
years later he wrote *Men Must Act* (New York: Harcourt, Brace, 1939), a
compelling polemic for U.S. entry into World War II.
[17]Christopher Lasch points out that in the early 1920's the tendency
among Mumford and his contemporaries toward cultural criticism often
reflected the mistaken assumption that a so-called cultural transformation
can take place apart from political change. See Christopher Lasch, "Lewis
Mumford and the Myth of the Machine," *Salamagundi* 49 (Summer 1980):
4–28.

down apartment on West 96th Street.[18] Poverty was not rapidly overcome. For much of his early adulthood, Mumford had to support his mother, making the burden of earning a living all the more difficult.

Mumford's social marginality and emotional uncertainties may have contributed to his sense of being dispossessed and to his determination to have an impact on the world. He lacked the established social position that facilitates assuming the office of the intellectual, and his academic training did not fundamentally alter this. When he entered New York's excellent school of science, Stuyvesant High, he studied electronics, with the intention of becoming a radio technician. His ambition to become a writer enlarged his aspirations, and when he graduated from high school after only three years, he enrolled at City College in 1912, excelling in English literature, politics, and philosophy.[19] Though he attended a large number of classes at City College and later at Columbia and New York universities and the New School for Social Research, he never took a degree.

Mumford's studies were determined more by his intellectual interests than by the stiff academic requirements of his day. Rather than following the prescribed course of study leading to a bachelor's degree, Mumford engaged the world of books and, more importantly, the city itself.[20] As a boy, he walked the city and visited museums with his grandfather. He delighted in the variety of street scenes in a city that still had open vegetable markets and views of the waterfront.[21] He considered himself "a child of the city": "New York," he claimed, "exerted a greater and more constant influence on me than did my own family."[22] As much as his classes, his own reading pro-

[18]Donald Miller, *Lewis Mumford: A Life* (New York: Weidenfeld & Nicolson, 1989), pp. 13–19.

[19]Ibid., pp. 50–51.

[20]Mumford, "First Decade," personal memoir, 15 July 1961, Mumford Papers.

[21]Miller, *Lewis Mumford*, pp. 26–27.

[22]Mumford, *Sketches from Life* (New York: Dial, 1982), pp. 3, 25; quoted in ibid., p. 3.

vided entry into the world of ideas. By the time he was twenty, he
had identified Patrick Geddes's, *Cities in Evolution*, George W. Russell's
Cooperation and Nationality, and Walt Whitman's *Leaves of Grass* as
the three books that most "altered my habits of thought and ways of
living." Geddes confirmed for Mumford the fundamental importance
of human interaction with the environment; Russell, a leader in the
Irish cooperative and rural revival movements, taught him the value
of folk culture; and Whitman demonstrated the fundamental importance of the creative imagination.[23] His early acquaintance with these
books demonstrates the diversity of Mumford's interests: nature and
society, art and science, literature and self.

Enlistment in the U.S. Navy in April 1918 disrupted his personal
life but also provided an important stimulus to his concern with public issues. After the war he resolved to make his living by writing,
putting aside boyhood dreams of becoming a playwright for the more
attainable goal of being a cultural critic. In 1919, at the age of twenty-
three, he was appointed associate editor of the *Dial*. But when the
magazine failed in its attempt to become a major journal of cultural
criticism and social reconstruction (a victim of the return of "nor-
malcy" in the postwar era), Mumford lost the most secure job he was
to have for more than two decades. He began to write for the *Freeman*, the *New Republic*, and the *Journal of the American Institute of
Architects* (AIA); and he published his first book, *The Story of Utopias*, in 1922. A year later he helped organize the Regional Planning
Association of America. He completed two important studies of
American culture: *Sticks and Stones* in 1924, a turning point in his
career, for the book was especially well received; and *The Golden Day*
in 1926. These were followed by *Herman Melville* (1929), *The Brown
Decades* (1931), and the first two volumes of the "Renewal of Life"
series, *Technics and Civilization* (1934) and *The Culture of Cities* (1938).
Beginning in the 1940s, various universities acknowledged Mumford
with honorary degrees and offers of visiting professorships. Royal-
ties on book sales and a byline in the *New Yorker* finally provided
Mumford with a comfortable living in the postwar era.

[23]Mumford, untitled, unpublished MS, 1916, Mumford Papers.

As a writer who set out to be what Russell Jacoby calls a "public intellectual,"[24] Mumford positioned himself as a mediator and transmitter of ideas. But in the process of transmission, Mumford created his own particular synthesis: an interdisciplinary approach that drew from developments in the social sciences and cultural criticism. Although he is best known as a student of cities, Mumford was a surprisingly comprehensive writer.[25] His books and articles cover topics from art, architecture, and literature to sociology and the history of technology. This range of interests reflects Mumford's holistic approach to the study of society and culture. Mumford united the study of technology and society with cultural studies in architecture and literature because he had become convinced at an early point in his career that an interdisciplinary approach could best address the crises that afflicted Western civilization in the modern era. Chief among these, he thought, was the increasing technological capacity for destruction of both the human and natural worlds. Mumford was not a technological determinist, nor was he inclined toward explicitly "political" explanations. He sought to understand the relationship between culture and the predominant technological and social patterns of American life. Trying to uncover the underlying values, beliefs—the weltanschauung of American culture—Mumford also hoped to recover an alternative American literary and artistic tradition.

Thomas Bender points out that Mumford sought to "fuse humanism and science, [thereby] becoming a sort of literary sociologist."[26] Not surprisingly, then, among the three men Mumford said were seminal influences on the development of his ideas, two were literary critics and the third a scientist: the botanist and geographer Patrick Geddes,

---

[24]Russell Jacoby, *The Last Intellectuals: American Culture in the Age of Academe* (New York: Farrar, Straus and Giroux, 1987), p. 5, defines *public intellectuals* as "writers and thinkers who address a general and educated audience."

[25]A number of recent books have included serious discussion of Mumford's work; among them is T. P. and A. C. Hughes, eds., *Lewis Mumford: Public Intellectual* (New York: Oxford University Press, 1990).

[26]Bender, *New York Intellect*, p. 235.

whom Mumford met on a trip to Britain in 1920, and the critics Randolph
Bourne and Van Wyck Brooks. Geddes, who with Victor Branford published the British journal *Sociological Review*, was a regionalist and planning advocate; his book *Cities in Evolution* (1912) turned Mumford "back
to the earth itself as the fundamental postulate."[27] Geddes's work in building a regionalist social science that blended sociology and geography—
"sociography"—seemed to Mumford an essential step in developing the
new science needed to deflect the tremendous destructive impact of
human development on the environment.

At the *Dial*, where he worked with Bourne and Thorsten Veblen,
Mumford was introduced to cultural criticism as a useful vehicle for
voicing dissatisfaction with American society and culture. Later, at
the *Freeman*, he worked for Van Wyck Brooks, the magazine's literary editor.[28] These journals, located in Greenwich Village, provided
"a special kind of sanctuary within the city itself from which to fly
the banners of revolt."[29] As a critic, Mumford was (as forever after-
ward) concerned with ethics as well as aesthetics: cultural criticism
was moral criticism. With other Young Intellectuals he espoused
cultural nationalism, disapproved of conventional bourgeois morality, and hoped for sweeping social change and institutional reform.[30]

Mumford distinguished his work by bringing cultural criticism to
bear on developments in geography and biology. This combination
of interests led him to the study of the city and its environs. The city
is in-between: it reflects both the dynamics of social relations and
human interaction with environment. Mumford's architectural criticism, already conspicuous by the mid-1920s in the *AIA Journal*, addressed the relevance of design to both.[31]

❖

In the early 1920s, Mumford came dangerously close to adopt-
ing the position that a cultural transformation of American life

[27]Mumford, untitled, unpublished MS, 1916, Mumford Papers.
[28]See Miller, *Lewis Mumford*, pp. 108–112, 149–150.
[29]Bender, *New York Intellect*, p. 230.
[30]Ibid., p. 231.
[31]Miller, *Lewis Mumford*, p. 171.

could occur without political and social reconstruction. The politics of the postwar era seemed so unforgiving, and his brief prewar attempt to find a political audience for his ideas was unsuccessful. Mumford had been a friend of Irwin Granich, who later (under the name of Michael Gold) published the Marxist journal the *New Masses*. Granich introduced Mumford to the alumni chapter of the Intercollegiate Socialist Society, which provided him with a forum for his first lecture, "Anarchism and Regionalism." But Mumford found that the audience, including Granich, largely ignored his idea of regionalism.[32] The experience undoubtedly reinforced his belief that politics must be superseded by a more fundamental transformation of values: change is "not merely a matter of appropriating catchwords or starting [political] parties: it is a matter of altering the entire basis upon which our present venal and mechanistic and life-denying civilization rests."[33] To some extent, it must also have prompted his rejection of Marxism as a mixture of "formulas and incantations."[34] His belief in cultural transformation was bolstered by his association with Van Wyck Brooks, who fostered Whitman's hope, expressed in *Democratic Vistas*, that literature could recover that lost "religious element."[35] Brooks, too, believed in the power of literature to revitalize the moral sense and renew the vitality of American culture. "The centre of gravity in American Affairs," Brooks wrote in 1915, "has shifted wholly from the plane of politics to the plane of psychology and morals."[36]

Mumford's interest in "psychology and morals" had first taken the form of support for feminism, which he saw as essential for lib-

[32]Ibid., p. 99.

[33]Mumford, "Reflections on Our Present Dilemmas" (1919?), published in *Findings and Keepings: Analects for an Autobiography* (New York: Harcourt, Brace, 1975), pp. 200–209.

[34]Mumford, "Second Decade," personal memoir, 16 July 1961; "The Year 1918," personal memoir, 1919; "Geography as a Basis for Social Reform," unpublished MS, 1917 (?), all in Mumford Papers.

[35]Walt Whitman, *Democratic Vistas*, in Mark Van Doren, ed., *The Portable Walt Whitman* (New York: Penguin, 1977), p. 337.

[36]Van Wyck Brooks, *America's Coming of Age* (New York: B. W. Huebsch, 1915); reprinted in *Three Essays on America* (New York: E. P. Dutton, 1934), p. 102.

eration from Victorian sexual mores. In 1914 he wrote an article arguing that women's liberation would be greatly advanced by the establishment of communal kitchens in apartment houses.[37] After the war Mumford directed his moral energy toward the conflict between civilization and nature. His growing fears about the course of civilization centered on the war. It occurred to him that progress meant the extension of science into the province of war: advanced technologies were applied for the purpose of systematic destruction. For Mumford, any adequate response to a crisis of civilization, marked by the attainment of the power to annihilate, required a fundamental change in thinking, a change that must begin in the psyche. For the destructive urges let loose by modern warfare and industry—which of course are the basis of the destruction of the natural world as well—rest on a limited notion of self, a definition of subjectivity rooted in opposition to nature. Mumford saw that changing the boundaries of self implied changes in culture, values, and worldview.

But what distinguished Mumford from Brooks was a sense that redefinition of self must accompany the recovery of civic participation. Thus community became essential to Mumford's work; he sensed that a revived democratic society was necessary to make regionalism an operative ideal, one capable of redefining economy and technology along ecological lines. These two projects—the investigation of the conception of self in American culture and the inquiry into the origins and prospects of regionalism—formed the basis of Mumford's work. The former led him to investigate the social and aesthetic implications of contrasting conceptions of subjectivity; the latter required the investigation of geography and the uses of new technologies and planning to effect a different kind of geographic order: the re-creation of a regional geography.

[37]On feminist households, see Mumford, "Community Cooking," *Forum*, July 1914, pp. 95–99. Mumford expressed disappointment with conventional strictures in "At Present," memorandum, 20 July 1919, Mumford Papers.

By developing a phenomenology of place, Mumford refashioned the Young Intellectuals' experimental linkage of culture and politics. The cultivation of the experience of place shaped Mumford's interest in regionalism as a cultural vision. And he connected this cultural project to a civic politics necessary to support regional planning.

CHAPTER 2

Defining Regionalism

During the 1920s and 1930s, regionalism was the focus of Mumford's work. He envisioned it as a new framework for modernity, for even as he searched the past for viable traditions to resurrect, his hopes were forward-looking. These hopes rested on the possibility of redirecting the technological legacy of the Enlightenment away from the accumulation of power over nature toward an ecological principle of interaction with the natural world. Regionalism was a principle that united three ideas: *"neotechnics"*—the adaptation of new technologies for the purpose of restoring the natural environment; *organicism*—the restoration of nature's influence on culture through literature, architecture, and the built environment; and *community*—the recovery of human-scaled, civic-minded social order.

While the foundation of Mumford's work rested on the scientific and literary implications of regionalism, the political and social aspects also engaged him. In two articles in the *Dial* in 1919, Mumford presented regionalism as a social theory that builds on the Enlightenment principles of democracy and self-government but goes beyond parliamentary liberalism to the restoration of civic democracy.

Mumford's response to the "belligerent nationalism," political repression, and widespread savagery of World War I was to posit regionalism against existing political structures. The disastrous results of the Treaty of Versailles confirmed his skepticism about the depth of the Western powers' commitment to democracy and national autonomy. For Mumford, Versailles meant that Wilsonian liberal internationalism had failed to distinguish itself as a genuine political alternative. In an article for the *Dial*, he reasons that inter-*nation*alism is predicated upon the nation-state, while meaningful political change must begin at the local and regional levels. Liberal internationalism rested on bureaucratic palliatives and was incapable of addressing the moral crisis of Western culture.[1] Seeking a radical alternative to conventional liberalism, Mumford thought that the movements for ethnic autonomy that proliferated after the war, especially in the former Austro-Hungarian empire, might be the beginning of a new urban and regional politics and political economy based on reclaiming a genuine civic life. And he associated regionalism and civic-mindedness with cultural diversity: regionalism would encourage cultural diversity and challenge the growing hegemony of an international metropolitan culture. "The future of nations," he wrote, "lies in the success which greets the efforts of *communities* and *associations* to establish corporate autonomy and to carry on their functions without subservience to that large and jealous corporation called the state."[2] The future, Mumford asserted, lay in new, cooperative, localist political and economic institutions that would revitalize the life of cities and regions. Here Bourne's vision of a culturally diverse and politically decentralized society was a clear influence.[3]

At the time Mumford did not consider the possibility that regional

[1]Mumford, "Wardom and the State," *Dial*, 4 October 1919, pp. 303–305. Less than twenty years later, however, Mumford invoked the tradition of liberal democracy in order to build opposition to fascism. See Chapter 1, pp. 14, and footnote 16, this chapter.

[2]Mumford, "The Status of the State," *Dial*, 26 July 1919, pp. 59–61, emphasis added; see also "The Status of Nationalities in the Great States," unpublished MS, August 1917, Mumford Papers, Van Pelt Library, University of Pennsylvania.

[3]See "Transnational America" *Atlantic Monthly* 118 (July 1916): 86–97; reprinted in C. Resek, eds., *War and the Intellectuals: Essays by Randolph S. Bourne* (New York: Harper & Row, 1964), pp. 107–123.

consciousness might encourage ethnic chauvinism and become a reactionary political tendency. Of course he opposed militarism and nationalism in all its forms and later condemned the Nazi ideology that justified a totalitarian social order by attaching place—the "soil"—to race. Mumford was interested in a progressive cultural regionalism, not an attempt to reverse the course of history. His political idea of regionalism involved the creation, or re-creation, of a civic culture and politics based on "compact and closely integrated communities."[4] But while he saw regionalism as a political possibility, he located its basis not in a political idea or movement but in the literary and artistic imagination. Regionalism concerns the imaginative recovery of place. This implied both a sense of place informed by the scientific and imaginative exploration of the environment and an idea of culture as linked to the geographic associations of place.

Recognizing Mumford's hope that regionalism could create a different social and political life is crucial to understanding the thrust of his work. The idea that alternative, cooperative institutions could be the basis for the "revival" of cities and regions is at the heart of Mumford's vision. But we also have to admit that Mumford failed to see the political uncertainties and difficulties this alternative raises. Thus he had little to say about the process of creating and sustaining alternative institutions and political practices, necessary to a democratic regionalism. His emphasis, rather, was toward developing a regionalist perspective grounded in a cultural and intellectual transformation, a new kind of humanism in the arts and sciences. By publicizing this perspective, one that looks differently at the relationships of nature to culture and of self to society, Mumford attempted to broaden the range of questions that define existing political discourses. And he resisted the tendency to make politics merely consequential. Thus the 1920s, when he became a principal of the Regional Planning Association of America, was the most politically active period in his life. His involvement with the planning and housing movements was a search for public openings, the political space into which he hoped to squeeze regionalist ideas and practices.

[4]See Mumford, "The Crisis of the Socialist Left," unpublished MS, 1915, Mumford Papers.

Mumford found an opportunity to study the scientific basis of a regionalist perspective when he accepted Victor Branford's invitation to go to London and become visiting editor of the *Sociological Review*. Mumford was attracted to the post because of his admiration for the work of the Scottish botanist, Geddes, Branford's associate. Branford, a financier turned sociologist, was an important contributor to the Sociological Society and to the *Review* at a time when the discipline of sociology was still unrecognized as a legitimate branch of study at British universities. Geddes began to write on the effect of geographic factors in social formations: during his informal study at the Sorbonne in Paris, he had become acquainted with the work of the French sociologist and geographer Frédéric Le Play. Like Geddes, Le Play was a biologist. Searching for a way to apply natural science to the study of society, Le Play proposed that human communities might be understood as extensions of the natural world, conceived of as a natural "region." He called the method of studying the interaction of all life within a region "sociography." Le Play promoted empirical studies of communities based on a triad of place, work, and folk. His work is considered a major influence in the revival of regionalism in the late 19th century.[5]

Geddes's appropriation of "sociography" framed his study of cities, culminating in his great work *Cities in Evolution*, published in 1915. Earlier, Geddes had applied Le Play's empirical method to an intensive "regional survey" of Edinburgh, but in composing *Cities in Evolution* he painted in broad strokes, considering both the existing state of cities and prescribing solutions to urban ills. One of his major concerns was the sprawling character of industrial (particularly British and North American) cities, a tendency he termed "conurbation." Geddes argued that the resulting destruction of compact and defined urban areas and the consequent destruction of the

[5]See Ronald Fletcher, *The Making of Sociology: A Study of Sociological Theory* (London: Michael Joseph, 1971), Vol. 2: 832–839; and Abbie Ziffren, "Biography of Patrick Geddes," in *Patrick Geddes: Spokesman for Man and the Environment*, (New Brunswick, N.J.: Rutgers University Press, 1972), pp. 3–12.

countryside could be countered only by the careful planning of urban expansion. In place of large, amorphous stretches of urbanized landscape, Geddes advocated the construction of carefully planned urban centers. He endorsed Ebenezer Howard's "garden cities": planned cities set in the countryside where population would be limited and adjacent land use would be tightly controlled to prevent sprawl.[6]

Geddes was attracted to the garden city idea because it provided a framework within which to reconstruct urban civilization. Because garden cities would be newly built, planners could take advantage of the technologies of the modern era. If the old metropolis reflected design based on the coal and steam era of industrialization ("paleotechnics"), the garden city would reveal the advantages of the new decentralized mode of production made possible by electricity ("neotechnics"). Geddes believed the era of livable cities, resource and scenic conservation, and civic consciousness lay ahead.

Mumford relied extensively on Geddes's characterization of the region and the city and on the importance he attached to techno-environmental factors in shaping cultural patterns. Geddes's attention to the garden city model and the regional survey were essential to Mumford's work. Indeed, the regional survey became the model for the most important document produced by the RPAA, the *Report of the New York State Commission of Housing and Regional Planning* authorized by the State of New York in 1926. In terms of the development of Mumford's regionalist perspective, Geddes's sociography provided something even more fundamental: a humanistic science of society that could be of use in the creation of communities grounded in a sense of place.

Geddes's sociology rested on the interaction of community, place, and work. As Robert Dickinson put it, he considered "the relations of these three as a mutual interaction in creating the spatial patterns of a society within a particular environmental setting."[7] The quality of human life could be understood in terms of the interaction of these variables in a particular place. Geddes argued that in its single-

[6]On Ebenezer Howard and the garden city movement, see Chapter 5.
[7]Robert Dickinson, *Regional Ecology: The Study of Man's Environment* (New York: John Wiley, 1970), p. 25.

minded pursuit of increased productivity, industrialization (as an application of new technologies) ignored its impact both on the environment and on social organization. The result has been imbalance between economy and ecology and between productivity and community. He thought science, a set of techniques based on empirical study, could propose remedies to these imbalances. For Mumford, the major lesson was the *instrumental* value of science in devising a functional response to the crisis of industrialized society. In particular, regional and community planning could provide a means of responding to the deteriorating conditions of social life. To use regional planning as an instrument for overcoming the disorder of the industrial landscape might be thought of as an exercise in foresight *and creativity*. A project of community and regional planning would be based on "a body of scientific investigation by which a human community may discover its movements and tendencies, evaluate its institutions, appraise its potentialities, and forecast its future development." In this sense, Mumford speculated, social science may be instrumental, for it seemed to have "come into its own" just at the point when the "basis of Western Civilization" had reached a crisis of "uncertainty" and paralysis.[8]

Under Geddes's influence, Mumford's idea of regionalism increasingly relied on the development of a new social science concerned primarily with regional structures, natural and built. Two conceptions are central here: the *geographic reality* of regional structures and the need for human activities to stand in some kind of *functional relation* to them. Mumford worked out these ideas in a series of articles on "regionalism" that he prepared for the *Sociological Review* and published in the late 1920s.

Mumford understood the region as a set of environmental relationships that extend over a certain area. Of prime importance are the geographical factors of terrain, climate, and soil that establish the conditions

[8]Mumford, "Sociology and Its Prospects in Great Britain," *Athenaeum*, 10 December 1920, pp. 815–816.

to which plant and animal life must respond.[9] Mumford insisted, there-fore, that the region is anything but ideal or abstract. Its existence re-sides in the "facts" of geography: climate, soil, and terrain constitute the "fundamental basis of existence."[10] The region sets the basic mate-rial conditions that underlie economic, technical, and social develop-ment. This idea reflects what Clarence Glacken identifies as one of the principal conceptions of the relationship between nature and culture in the history of Western thought: the concept of environment. In this sense nature lies outside social life and by virtue of its externality ex-erts instrumental power over human existence: culture must adapt to environment.[11] For Mumford, this does not mean that environment acts mechanically or routinely on the patterns of human life. As he put it, "The environment does not act directly upon man: it acts rather by con-ditioning the kinds of work and activity that are possible in a region."[12] Geographic conditions thus shape culture.

This idea is apparent in Mumford's discussion of the pre-indus-trial era, when "eotechnics"—the technologies and techniques that rely on local sources of energy—structured the sociocultural life of Europe. The design of houses, the use of indigenous materials, the layout of streets, the reliance on local sources of power such as wind and water all reflect a more intimate and functional relation to geo-graphic conditions. The material culture of pre-industrial Europe reflected the forces of environment; consequently, regional variation was pronounced. This meant beautiful *and* functional architecture and crafts—vernacular traditions that Mumford appreciated.

But Mumford's concerns were not solely aesthetic. He wished to shape the modern era by bringing forward regionalism as a perspec-

[9]Mumford, "Regionalism and Irregionalism," *Sociological Review* 19 (Oc-tober 1927): 277–288.

[10]Indeed, Mumford was so taken by the factual that for a time he thought of himself as a hard-nosed sociologist engaged in reviving Auguste Comte's search for a positive and "objective" science of human society. Mumford to Victor Branford, 1919, Mumford Papers.

[11]Clarence J. Glacken, *Traces on the Rhodian Shore: Nature and Culture in Western Thought from Ancient Times to the End of the 18th Century* (Berkeley: University of California Press, 1967).

[12]Mumford, "Regionalism and Irregionalism," p. 285.

tive appropriate to that era. For this reason, Mumford argued that the region remains a constant geographic influence on culture and society. He noted that even after the industrial revolution,

> The region . . . conditions economic activity. In the basic industries, as distinguished from the derivative industries, these conditions still have a rigorous hold. Each region has its basic economic complex. . . . It is absurd to speak of "industry" or the "machine system" as if the conditions that underlie its success, and the problems it faced, were the same in every region.[13]

"Basic industries" derive from the particular geographic advantages of a region, including available natural resources and relative accessibility. Even under a global capitalism determined to force geographical particularities into a *system* of economic specialization, the region remains important. Natural regions, Mumford argued, constitute their own systems, combining the power of climate, soil, hydrology, and flora to profoundly effect even the most powerful human structures. Admittedly, economic globalization has occurred in a manner that has very often undermined our potential for working with natural geographies. According to Mumford, the extension of markets and the perfection of production "are dead set against geographic realities." The universalizing tendencies of global capitalism, especially as magnified by the powers of industrial production, result not only in standardized products but a standardized commercial culture. Mumford called this "metropolitanism." Like the industrialized commercial cities it creates, metropolitan culture celebrates its capacity to overcome geography, and it is this kind of thinking that denies the fundamental symbolic and ecological significance of regional particularities. The abuse of land and resources is merely the consequence of our obliviousness. "We have disregarded the fundamental basis of existence," Mumford wrote, by treating "the land itself as if it were a vague shadow cast by growing cities." Rampant and unsustainable urbanization is an elemental and increasing threat to regional systems. There are limits, Mumford asserted, to nature's capacity to absorb exploitation. Overpopulation, particularly

[13]Ibid.

the concentration of people in huge metropolitan centers, creates an imbalance between human activities and "regional [ecological] realities"; how can the huge population of the metropolis meet its requirements for necessities such as water and food and disposal of wastes without spilling over into other regions and ruthlessly "exploiting" the resources of these less-developed areas? As examples Mumford cited the quest for water for Los Angeles and Chicago's sewage disposal problem. The "technological dodge"—an ever more extensive water-delivery system for Los Angeles or more elaborate sewage treatment facilities for Chicago—cannot indefinitely forestall the city's "inevitable dependence upon rainfall and catchment areas and forest reserves; sooner or later these conditions must bring all the plans for windy growth and illimitable land-speculation *down to earth*."[14]

Mumford's writings invite us to examine the choices we must make in determining our relation to the natural world. He saw this first and fundamentally as a moral choice between cooperation and exploitation, nurture and annihilation. However, Mumford's regionalism is not meant to produce a static moralism but to suggest a new set of principles to guide our engagement with nature. Regionalism expresses the hope that balances between human activities and "regional realities" can be found. The balance he advocated was meant to address the particular problems of an *urbanizing* society. He did not posit a static conception of balance, nor did he contemplate a return to pre-industrial society. Instead his conception of regionalism encompasses a wide range of human activities, including agriculture and industry, and requires the discovery of the potential that "each region has [for attaining] . . . a natural balance of population and resources and manufactures, as well as of vegetation and animal life."[15]

This should not be read as a rigidly teleological statement, as if a region had within it a single and transparent design for human usage. Nor is ecological regionalism to be perceived simply as an extension of natural design; regional ecologies must be considered in interaction with human activities. Human purposes cannot be entirely derived from nature. But while ecological regionalism must reflect and grapple with the social world, we cannot afford to ignore

[14]Ibid., p. 278.
[15]Ibid., p. 283.

the "geographic realities" of the natural world we inhabit. It is increasingly impossible to deny that human action determines the survival of the natural environment. This engagement with nature rests on the power of "man as geographic agent."[16]

Mumford found this idea in George Marsh's *Man and Nature* (1864), a study of the effects of human activities on the environment. Marsh understood the disastrous consequences that many human actions— even those that seemed benign—have had on the viability of natural systems. North Africa, for example, one of the bread baskets of the Roman Empire, was reduced to desert by overtillage, and the clearing of the forests of Tuscany choked its rivers with silt that filled harbors and created malarial swamps along the coast, driving away the human populations. But Marsh also noted that nature tends toward a "harmony" that produces an "almost unchanging permanence of form, outline and proportion." Should this natural and locally particularized "stability" be upset, whether by human intrusions or "geologic convulsions," nature will go about the restoration of its balance. This natural tendency toward a state of balance is usually at odds with human endeavor.[17]

Mumford agreed with Marsh that human activities interfere with natural balances, but he considered nature not as a permanent form but as a set of related biological and geographical processes. These processes, never static, produce a field of interaction. Plants, animals, and human beings interact with climate, soil, and other conditions to create an interdependent community of life. Consequently, while Marsh and other conservationists were well aware that human activity is an imposition on nature and argued for efforts to minimize and control its deleterious effects, Mumford looked forward to the possibility that in a new era, human activities could actually contribute to and participate in the *restoration* of the community of life. Accordingly, Mumford believed that human activity might be undertaken with the intent to aid nature. He regarded the region as the basis for enacting this recuperative process, because the region is a

[16]This is another major principle of Western thinking about nature, as identified by Glacken in *Traces on the Rhodian Shore*.

[17]George Perkins Marsh, *The Earth as Modified by Human Action* (New York: Charles Scribner's Sons, 1885) p. 16, emphasis added. *The Earth* was a revision of *Man and Nature*.

workable field in which to study and plan human interaction with the environment.

As a field, the region is related to the idea of an ecosystem, which the botanist Frederick Clements defined as "a unified mechanism" whose "whole is greater than the sum of its parts." The commonality and functional interdependence of the plant and animal communities is a "biotic formation" or "biome": "an organic unit comprising all the species of plants and animals at home in a particular habitat."[18] For Mumford, the biological biome is part of the geographic region. Unlike Clements, Mumford emphasized the specifically geographical rather than biological, qualities of this larger field. These larger natural fields provide a conceptual framework for bringing human activities into relation to nature. Furthermore, the region, like the biome, establishes a set of functional relationships within the field that makes it possible to consider the consequences of human activities. The implication of this is clear: we are not so much bound to a law or order of nature as *engaged* in a reciprocal relationship with the natural world. This dynamic of reciprocity makes for the complexity of the human-natural nexus, a reciprocity that for Mumford is not superseded even by the most intensive human alteration of the environment. Hence, urban regions are at the same time natural regions, for natural influences never disappear even in areas subjected to the greatest human usage. "Even in its most highly developed stages," Mumford wrote later in *The Culture of Cities*, "the city is, among other things, an earth form. It is put together out of wood, stone, clay, asphaltum, glass. Its shape is conditioned by topography and the nature of the land; and the special requisites of the site."[19]

The practical interaction between human beings and the natural world encompassed by the region suggested to Mumford that the region is both "natural" and "cultural." The "natural region" is rightly

[18]Clements quoted in Donald Worster, *Nature's Economy: A History of Ecological Thought* (Cambridge: Cambridge University Press, 1977), p. 20. Also see Frederick Clements, *Bio-Ecology* (New York: John Wiley, 1939), p. 7.

[19]Mumford, *The Culture of Cities* (New York: Harcourt, Brace, 1938), p. 316; hereinafter referred to as *C of C*.

the object of scientific inquiry. The study of the region provides systematic knowledge of the interaction of all beings in the geographic field: the relationships of climate and topography and biota. A region is also "cultural," an elaboration of human experience (past and present) grounded in place.

Mumford's cultural regionalism parallels what Donald Worster calls the "arcadian tradition." Worster's characterization of 18th-century British nature essayists applies equally well to Mumford:

> What is important is the need felt by these nature essayists to locate a compelling image of an alternative world and an alternative science. That, after all, is the principal function of myth: not to establish facts but to create powerful symbols and designs that can explain the inner core of human experience and provide dreams to live by.[20]

For Mumford, the "powerful symbol" that explained "the inner core of human experience" was regionalism. Regional consciousness, a cultural elaboration of the sense of place, makes possible a new dialogue with the natural world. In his view this cultural elaboration of place created a *political context* for the instrumental use of science in investigating and reshaping human interaction with the region. Thus Mumford brought together the spheres of science and humanities. He realized that the application of science cannot in itself address the crisis of civilization nor right the imbalance between the human and natural worlds. The potential instrumental value of science and technology is important, but this should not obscure the danger of seeing the progressive unfolding of the logic of technical "reason". Science itself must be weighed by a literary and cultural criticism aimed at restoring the idea of an organicist, regional culture.

---

[20]Worster, *Nature's Economy*, p. 77.

CHAPTER 3

Community and Place

Returning from England, Mumford began a survey of New York and its environs modeled on Geddes's technique of the regional survey.[1] Mumford was forced to recognize the accelerating destruction of the fields, marshes, and beaches surrounding the city. He realized that this was a new kind of assault on nature, not a recurrence of the 19th-century onslaught of mines and smokestacks. This was an "invasion" driven by the desire to consume nature. It was the kind of invasion he witnessed on Staten Island, where "a thousand people who liked sunlight and salt air had purchased building-sites and had put up bungalows and had devastated a strip of beach."[2]

"Liking sunshine and salt air" may reflect an instinctual drive for life, but in this case instinct was recast into a culturally defined desire for things. Like Thorstein Veblen, Mumford understood that

[1]For a description of Patrick Geddes's "regional survey" method, see Robert Dickinson, *Regional Ecology: The Study of Man's Environment* (New York: John Wiley, 1970), pp. 24–25.

[2]Mumford, "Environmental Degradation," in *Findings and Keepings: Analects for an Autobiography* (New York: Harcourt, Brace, 1975), pp. 63–64.

habits of waste and "conspicuous consumption" constituted a new ethos that feeds the escapism, "pleasure-seeking," and pursuit of status characteristic of mass culture.[3] This ethos was the psychic basis for consumer culture and the justification for a new kind of exploitation of the natural world. Particularly in the emerging metropolitan areas, the potential for destroying nature had been greatly enhanced. As the city spread outward into the surrounding countryside, as adjacent lands were given over to the "recreation" of the urban dweller, as the older working landscapes were destroyed, and as beaches and mountains became "resorts," a new exploitation of nature became firmly established. It did not displace the exploitation so characteristic of the 19th century—the mining of natural resources for industrial production (the kind of economy that wrecked entire regions of Appalachia); it simply added to it the consumption of nature as amenity. An urban culture's embrace of the natural world as a form of recreation actually trivialized the natural world. In short, nature was made a mere backdrop to the culture of the metropolis.

While Mumford was concerned with planning an ecological and regional restructuring of the city and its environs, he insisted that regional planning be informed cultural criticism. For Mumford, culture was a worldview, an outlook on life that encompassed personality and society. His view of culture was related to what the anthropologist Robert Redfield calls "ethos": a "group personality" reflective of the "whole community" and linked to the "whole inner structure of the person."[4] In his cultural criticism Mumford examined the relationship between subjectivity (the "inner structure" of self) and the aesthetic and social forms of the "whole community."

[3]Later, in *C of C*, p. 258, Mumford characterized mass or "metropolitan" culture as derivative, lacking firsthand experience: it is "a world where the great masses of people, unable to have direct contact with more satisfying means of living, take life vicariously, as readers, spectators, passive observers."

[4]The anthropologist Robert Redfield developed the conception of culture as worldview and ethos. A cognitive anthropologist, Redfield considered the "mental states of other people [as] the very stuff of . . . cultural studies." See Robert Redfield, *The Little Community and Peasant Society and Culture* (Chicago: University of Chicago Press, 1960), esp. pp. 86, 88.

But he was just as interested in finding discontinuity, the "false starts" of the American experience, as he was in tracing the ethos that accounts for the historical continuities of American cultural development. Thus, in his first two cultural studies, *Sticks and Stones* (1924) and *The Golden Day* (1926), he employed both archaeological and historical approaches. He wrote a history of American culture, tracing its intellectual sources and uncovering its popular mythologies, and he attempted an archaeology of American origin: he looked for the remnants, the ruins of an alternative American culture. His subject was the relation between culture and nature, and he found in the ruins of the culture of provincial New England's "golden day" the remnant of the America of "settlement" that had articulated a sense of place—a culture that he contrasted with the historical America of movement and exploitation. His purpose was to present an earlier America that for the modern observer could stand as an imaginative restoration of nature. This America of the "golden day" constituted a "usable past," a kind of myth that presents the possibility of reconceiving the self and reframing culture.

Mumford hoped to define a subjectivity inclusive of the sense of place. He drew from two sources: one was romantic and concerned the value of spontaneity and aesthetic contemplation; the other was functionalist and advanced the intrinsic value of productive work. In this second regard Veblen was an important influence.[5] Mumford shared Veblen's appreciation of the "instinct for workmanship," recognizing it as an important human capacity. For Veblen, the constructive work of producing needed things was a necessary discipline and an essential part of the opposition to the capitalist ethos that fosters the self as consumer. This ethos, according to Veblen, finds its highest expression in a "business class" that values predation over production, "conspicuous consumption" over utility. Like John Ruskin, Veblen saw in pre-industrial crafts the embodiment of the

[5]Mumford and Veblen both worked at the *Dial* after the war. On Veblen, see Joseph Dorfman, *Thorstein Veblen* (New York: Viking, 1955), esp. pp. 411, 418; John P. Diggins, *The Bard of Savagery* (New York: Seabury, 1978).

"instinct for workmanship"; but unlike Ruskin, Veblen recognized that this instinct survived in the modern era: the craftsman has been succeeded by the engineer, who has become the repository of the values of care, productivity, and meaningful effort.

Most important here is not work, per se, which, after all, is socially constructed, but the notion of livelihood: the labor that sustains a meaningful life. The belief that dignity is to be found in identifying with the "means of livelihood" links Veblen and Mumford to Paul Goodman, whose *Growing Up Absurd* (1959) criticizes the culture of the young because it rejects the productivity and (potential) creativity of work. Like Veblen and Goodman, Mumford believed that productive engagement is crucial to self-realization. For livelihood is a way of being-in-the-world, linking the individual to its materials and inhabitants—overcoming alienation.

As a social scientist, however, Veblen did not divorce a theoretical or experiential discussion of livelihood from the economic issues of the early 20th century. And in *The Engineers and the Price System*, he linked the social value of the "instinct for workmanship" with the economic value of "efficiency": the key to change, he asserted, lay with the new technical class, organized as a "soviet of technicians." Only these technocrats could operate the industrial system efficiently and in the interest of the public good.[6] In one important respect, this analysis connects Veblen to the mainstream of Progressive Era thinking: Progressives used the idea of efficiency in production methods to justify (even sanctify) the modernization of industry. And efficiency meant both the introduction of mass production technologies and the "scientific management" of work and workers by engineers along lines suggested by Frederick Taylor.[7]

Mumford was uncomfortable with Veblen's unqualified endorsement of the canon of efficiency and responded to it in an article entitled "If Engineers Were Kings," published in the Freeman in 1921.

[6]Thorstein Veblen, *The Engineers and the Price System* (New York: B. W. Huebsch, 1921).

[7]For a good critique of Taylorism, see David Montgomery, *Workers' Control in America* (Cambridge: Cambridge University Press, 1979); and Herbert G. Gutman, *Work, Culture and Society in Industrializing America* (New York: Knopf, 1976).

He put his critique simply: "What [does it matter] if industrial society is run more efficiently, if it is run in the same blind alley in which humanity finds itself today"[8] Mumford valued productivity and utility, but he dislikes a technological culture, a culture that prizes productivity to the exclusion of everything else. Consequently, he broke with Veblen in this respect: the (subjective) value of workmanship, like the social value of productivity, must find a larger referent, some greater end or purpose. For work to be meaningful, for it to renew individual (and cultural) creativity, it must become a livelihood grounded in a broader understanding of life.

Mumford's critique is aesthetic and moral, not political; it is important because he keeps his attention on the self and on the connection between self and nature.[9] And it helped him to reintroduce his concern with art and literature as necessary to the imaginative perception of life: an understanding needed for the revival of the inner life.

Like Harold Stearns, who celebrated instinct—an "America of our natural affections"—Mumford wanted to reclaim the earlier America as he perceived it, the America that fostered the inner life.[10] But he

[8]Mumford, "If Engineers Were Kings," *Freeman*, 23 November 1921, pp. 261–262.

[9]It is also a defining moment in his work. For Mumford did not pursue the political critique of Veblen along the lines suggested by Petr Kropotkin in *Fields, Factories and Workshops; or, Industry Combined with Agriculture and Brain Work with Manual Work* (New Brunswick, N.J.: Transaction Publishers, 1993; reprint of London: T. Nelson, 1912), a work that Mumford admired. Kropotkin's advocacy of a revolution in the organization of production based on the development of small-scale, technologically advanced workshops under the control of the workers might be read as a radical elaboration of Veblen's principle of "workmanship" at the same time that it is a critique of the technocratic elements of Veblen's work.

[10]Mumford and Stearns collaborated on *Civilization in the United States* (New York: Harcourt, Brace, 1922). The quotations are taken from Stearns's contribution, "The Intellectual Life." Stearns was influenced by Bourne's linking of cultural and generational change: Randolph Bourne, *Youth and Life* (Freeport, N.Y.: Books for Libraries Press, 1967; orig. pub. 1913).

also feared that the search for instinct would degenerate into a quest for mere individual "enjoyment and satisfaction." Mumford argued that repossessing the romantic emphasis on freedom, spontaneity, and love must not rest on a false opposition to work, productivity, and practicality.[11] Determined to avoid the path of romantic isolation and retreat, Mumford wanted to build on but not be limited by what F. O. Matthiessen called the "oppositional self." Mumford posited this romantic tradition of opposition to society and connection to nature as a way of reintroducing organicism. Still, his hope was to define a subjectivity that embraced the natural world without negating the importance of practical utility or a functional social order. An aesthetic point of view, then, became a part of his larger moral vision of the social and natural worlds.

In *The Golden Day: A Study in American Literature and Culture*, Mumford found a moral-aesthetic vision in Emerson's *Nature* (1836). Emerson's opening to nature was founded on his (re)conception of "experience": he reconsidered the "established hierarchy of experiences" that fixed the relationship of human to God—a cultural universe riveted to the Word. In doing this, Emerson also proposed a "democratic challenge" to the "old lines" of experience and thought that "were dying out."[12] Emerson went not to accounts of nature but to nature itself, that is, directly to experience.

By shifting the focus of interpretation from the idealism of the Transcendentalists to their organicism, Mumford presented an im-

[11]In "The Collapse of Tomorrow," *Freeman*, 13 July 1921, pp. 414–415, Mumford was writing in response to Stearns's "Intellectual Life." Stearns clearly favored art over business, spontaneity over intention, nature over technology. He also implied that cultural transformation could overcome the institutional (business-oriented) structures of American life.

[12]Mumford, *The Golden Day: A Study of American Literature and Culture* (New York: Boni and Liveright, 1926), p. 97; hereafter referred to as *GD*.

portant and original interpretation.[13] He discovered a neglected side of Emerson's thought: nature not as a moral and aesthetic absolute but as a process of growth that sustains our faith in the possibilities of life. In his essay "Circles" Emerson wrote that "there are no fixtures in nature. The universe is fluid and volatile." This fluidity is a process of decay and destruction, rebirth and growth manifested in natural and human endeavors: "The new continents are built out of the ruins of an old planet; the new races fed out of the decomposition of the foregoing. New arts destroy the old . . . the investment of capital in aqueducts made useless by hydraulics; fortifications, by gunpowder; roads and canals, by railways; sails, by steam; steam by electricity."[14] Nature is active, a process of change and growth that mirrors human growth and may become a source of experience. In this sense nature is neither an ideal standing above culture from which all values derive nor an outer realm of natural laws understood by science as the external conditions that limit and define all life. Rather, nature may be understood as processes of life, creative and destructive, always active, that unite us with the universe.

This is the spirit that Mumford reclaimed from the Transcenden-

[13]Mumford's work broke ground for F. O. Matthiessen's *American Renaissance: Art and Expression in the Age of Emerson and Whitman* (New York: Oxford University Press, 1941). At the same time, Mumford did not acknowledge that Emerson's theory of knowledge was dualistic; Emerson addressed consciousness and matter through two different human faculties, the "Understanding" and the "Reason." Mumford downplayed this, in effect simply ignoring Emerson's philosophical idealism. Another important reading of Emerson is Sherman Paul, *Emerson's Angle of Vision* (Cambridge: Harvard University Press, 1952), esp. p. 72. Self and universe, consciousness and nature, mind and body are not reducible to the other; they are parallel or in "correspondence" and related through living relations, the organic principle: "Nature says—he is my creature, and . . . he shall be glad with me. Not the sun or summer alone, but every hour and season yields its tribute of delight; for every hour and change *corresponds* to and authorizes a different state of mind." Correspondence suggests the "spiritual mediation of nature" by positing the unity of the finite (outer world) and infinite (inner world).

[14]Ralph Waldo Emerson, "Circles," in William G. Gilman, Ed., *Selected Writings* (New York: Signet, 1965), p. 296.

talist traditions: infusing self into the world and combining the practical with the intuitive. Emerson's work exemplifies the spirit of pragmatic engagement with the world. Mumford quoted Emerson approvingly: "'nothing is at last sacred but the integrity of your own mind'" (*GD*, p. 97). The crucial thing is Emerson's "intellectual, or cultural, nakedness: the virtue of getting beyond the institution, the habit, the ritual, and finding out what it means afresh in one's own consciousness" (*GD*, p. 96). Mumford considered this practice of "self-reliance" a means of finding an authentic relation to the world through experience.

❖

The Golden Day presents the question of American cultural development as "the challenge of the new American society," a challenge created when "the European [emigrant] lost the security of his past in order to gain a better stake in his future" (*GD*, p. 96). This trial, according to Mumford, requires neither the rejection of modernity nor its blind acceptance. Mumford rejected both sides of the prevailing cultural debate: a Eurocentric romanticism that annihilated modernity by reviving the memory of the ancien régime and a modernism that accommodated the "megamachine" by deriving its aesthetic sensibility and spatial form from the new technologies. For Mumford, this unsatisfactory polarity in the cultural discourse mirrored an equally inadequate political discourse between collectivism (socialism) and individualism (reactionary capitalism). He saw the failings of culture and politics reflected in the crisis of modern, centralized organization: its periodic crises of overproduction, its tendency to waste and inefficiency, its disastrous environmental effects, its unwieldy and disagreeable cities.

Rethinking American culture would be greatly enhanced by presenting a narrative of America's "usable past," an idea Van Wyck Brooks suggested in his 1918 *Dial* essay. By "usable" Brooks meant relevant to the spiritual and critical needs of the present.[15] Noting the postwar malaise that afflicted the young, Mumford argued that

[15]Van Wyck Brooks, "On Creating a Usable Past," *Dial*, April 1918, pp. 337–341.

a reconstructive critique of culture was necessary to clarify his generation's "special relations with its past." By uncovering alternative cultural paths, his generation could "gain the ability to select qualities which it values."[16] In antebellum New England Mumford found a "usable past" that could serve as a source of values, a reserve of the imagination. *The Golden Day* is not an exercise in nostalgia: "we cannot return to the America of the Golden Day, nor keep it fixed in the postures it once naturally assumed; and we should be far from the spirit of Emerson or Whitman if we attempted to do this" (*GD*, p. 278). Rather, the golden day can serve as a useful myth, a necessary criticism of a modern culture that exposes the limitations of a "machine civilization."

Mumford derived many of his themes from Brooks whose essay *Letters and Leadership* (1918) reveals a culture much too preoccupied with grand schemes for "improvement" and "development." The Americans, Brooks wrote, have failed to cultivate "life for its own sake" and are incapable of "cooperating with nature." Brooks decried the "pioneer mentality" that has "robbed and despoiled and wasted [nature] for the sake of private and temporary gains"; he observed the results of this mentality in a landscape littered with "the wastes and ashes of pioneering." Brooks attributed this pioneer mentality to Puritanism, which "by making human nature contemptible and putting to shame the charms of life . . . unleashed the acquisitive instincts of men."[17]

Mumford also owed much to Brooks's method: the analysis of a culture by means of ideal types, social mores, values, and worldview. But whereas Brooks argued that unlike the European, the American lacked a mythology and was given to material acquisitiveness in the

[16]Mumford, "The Collapse of Tomorrow: The Emergence of a Past," *New Republic*, 25 November 1925, pp. 18–19.

[17]Van Wyck Brooks, *Letters and Leadership* (New York: B. W. Huebsch, 1918), reprinted in *Three Essays on America* (New York: E. P. Dutton, 1934), esp. pp. 116–118, 131, 134.

service of reason and practicality[18], Mumford contended that America was very European: America, he maintained, was an outgrowth of a cultural process that began in Europe with the "disintegration" of the medieval culture and the twin revolutions of Western thought represented by the Enlightenment and the Reformation. This intellectual inheritance derives from a "process of abstraction . . . an attempt to isolate, deform, and remove historic connections" (*GD*, p. 25).[19] In the resulting worldview, the universe is understood in mechanistic terms and the culture it generates, "having no past, and no continuity," has "no future" (*GD*, p. 178). No wonder the pioneer (understood as an ideal type) failed to consider his connection with nature and gave himself over to the "utilitarian conquest of . . . [the natural] environment" (*GD*, p. 42). Mumford understood that this utilitarian worldview, though founded on the role of reason and furthered by "abstract thinking," developed its own mythology. It created an ethos of independence and self-reliance, the basis of frontier capitalism.

Thus *The Golden Day* also presents Mumford's evidence for an "unusable" past, a past that Mumford located in the central myth of the American experience: the "*Natur-Mensch*," first given expression in James Fenimore Cooper's Leather-Stocking tales. Natty Bumppo was "versed in the art of the woods, with the training of the aborigine himself; he shared the reticence and shyness that the Amerind perhaps showed in the company of strangers; and above the tender heart . . . he disclosed a leathery imperturbability" (*GD*, p. 64). Mumford traced the origin of this mythic type back to Jean-Jacques Rousseau's atavistic vision of life. In spirit and principle the romantic movement rejected the Enlightenment belief in "progress, science, laws, education, and comfort": a worldview in which "progress was the

[18]In Brooks's first published work, *The Wine of the Puritans* (London: n.p., 1908), reprinted in C. Sprague, ed., *Van Wyck Brooks* (New York: Harper & Row, 1969), he had contrasted the "purely intelligent"—rationalistic—North American with the organic European. Europeans are "molded by the special traits of climate, natural elements and properties of their lands." The problem was the Americans' failure to embrace all of life.

[19]According to Marshall Berman, *All That Is Solid Melts into Air* (New York: Penguin, 1982), p. 15, modernity "pours us all into a maelstrom of perpetual disintegration."

mode and comfort the end of every civil arrangement" (GD, p. 54). But while the pioneer substituted nature for progress, imitating the romantic quest for "a clear sky and open fields," "he wanted comfort no less" (GD, p. 54). In the end the legacy of the romantic spirit to the American character was largely negative: Americans accepted the premise that nature should be sought by "escape[ing] the conventions of society." This meant that even in admiring the natural world, even in embracing nature, Americans would evade responsibility for it. Worse, the pioneer found it convenient to separate principle from practice. And while embracing nature as an escape from social evil, he "made no bones about appealing to the Central Government when he wanted inland waterways and roads and help in exterminating the Indian" (GD, p. 54). In the end "society was effete," and at the same time "its machinery could be perfected—the pioneer accepted both notions" (GD, pp. 54-55). The result, according to Mumford, was the perfect engine of "destruction and pillage." The frontiersmen lacked the "peasant habits or the ideas of an old culture," and the impact of their destructive practices was magnified by the proexpanstionist policies of the federal government (GD, p. 58). Under these conditions medieval European culture unraveled. The only thing distinctive about the American culture that replaced it was its rootlessness. Mumford's frontier is a place where "the civil man could become a hardy savage, where the social man could become an `individual,' where the settled man could become a nomad, and the family man could forget his old connections" (GD, p. 57).

Mumford's oversimplification of the history of the westward movement can be accepted only as a polemic that ignores much of the significance of the frontier in American history. Mumford said nothing of pioneers as explorers or of the social dimensions of the frontier experience.[20] Yet his argument can be appreciated for its moral

[20]By contrast, William Carlos Williams, in his imaginative history of America, *In the American Grain* (New York: New Directions, 1956; orig. pub. 1925), pp. 130–39, suggests the importance and excitement of exploration. Daniel Boone represents the pioneer as explorer, and his exploration is an ecstatic "devotion to the wilderness." The behavior of the pioneers must also be understood in social terms. Western settlement represented a significant opportunity for the unpropertied to escape oppressive social conditions.

critique of American civilization—the selling out to "dreams of a great fortune in real-estate, rubber, or oil" (*GD*, p. 69). Mumford showed that such a culture, which depends on commerce and commodity, derives in some part from the pioneers' "warfare against nature" (*GD*, p. 59).

There is a direct passage from "the chief pioneering experiences" of hunting and trapping to those of American capitalism: "it required only a generation or two before the trader [of furs] became the boomtown manufacturer, and the manufacturer the realtor and financier" (*GD*, p. 65). For Mumford, American capitalism was built upon and appropriated the frontier mythology. The individualistic assertion of power represented by the laissez-faire or free market ideology is the urban-industrial equivalent of pioneering. They share a psychic principle: the (false) hope of transcendence through conquest. The pioneer's conquest of the land became the "inventor-businessman's search for power" (*GD*, pp. 43, 73–74).

Mumford's concern, of course, was with the present: with the habits of mind that prevent cultural change. He was convinced that as long as this pioneer mythology went unchallenged, America would be destined to experience spiritual impoverishment and undergo a technological holocaust of the natural world. Having unleashed the longing for acquisition of things, America would lead the world into an era when all values—human and natural—are subsumed in a mad rush to create (private) wealth. For Mumford, the pioneer mythology embodied the destructive consequences of the revolutions in thought that created modernity. However, his purpose was not to negate modernity but to recover an alternative based on integrating the Enlightenment principle of technical progress and scientific inquiry with the principles of civic democracy and a religious reverence for nature. In effect, Mumford built his search for an alternate modernity on his critique of the pioneer mentality.

▦

The Golden Day concerns the conquest of space by a European-derived culture. Commenting on *Moby Dick*, Mumford pointed out that the tendency to spatialize was connected with the desire to conquer and exploit: "the Whale is Nature, the Nature man warily hunts and studies, the Nature he captures, tethers to his ship, cuts apart, scientifically analyzes, melts down, uses for light and nourishment, sells in the market." (*GD*, p. 149). One is reminded of the poet Charles Olson's representation of the American character in his study of *Moby Dick*. Having himself "lived intensely his people's wrong, their guilt," Melville disclosed the American tragedy, the resolve to "act big" and consequently to "misuse our land, [and] ourselves." By "reaching back through time until he got history pushed back so far he turned time into space," Melville created in *Moby Dick* a mythic stage for the reenactment of the central American drama: the conquest of a continent. Since Ahab "had all space concentrated in the form of a whale," the battle with the whale is the battle with a continent. Though historical, this conflict is at the same time spatial; it demands an end, conquest, a goal that repudiates the earlier American hope of finding/creating place—a hope of origin that Mumford located in the early America of "settlement" that preceded the culture of the pioneer. As Olson reminds us, "to Melville it was not the battle to be free but the will to overwhelm nature that lies at the bottom of us as individuals and a people."[21]

Mumford understood that having a pioneer culture resulted in our failure to experience the fullness of the American continent, a failure to realize the particularity of regions. For the 19th-century observer, the frontier was a game marked by a steadily advancing "line" of conquest; for the pioneer the continent was big and empty. In the 20th century this empty space becomes the "raw land" of the real estate developer: it exists to be conquered or consumed. Recovering nature begins with the imaginative turning of empty space into natural space. Mumford was aware of this as a part of the process of cultural transformation: the empty space of the continent may be reimagined as the natural space of the region, the ethos of conquest

[21]Charles Olson, *Call Me Ishmael* (San Francisco: City Lights, 1947), pp. 12, 14–15. My reading of Olson benefits from Sherman Paul, *Olson's Push* (Baton Rouge: LSU Press, 1978); on topos, see p. 33.

and exploitation may give way to the ethos of recovery and restoration. For Mumford, this cultural process of transformation rested on the cultivation of a self-reliance, a subjectivity rooted in the recovery of the geographical relationships of place. The conception of natural space refers both to the external world of the natural region and to the interior world of self. Place unites both: it requires not only the recognition of the natural ecology of the region but an imaginative opening of inner space. Only the cultivation of place will stand *against* the culture's obsession with linear time and the conquest of empty space.

Even before he wrote *The Golden Day*, Mumford's first important book, *Sticks and Stones* (1924), addressed the problem of turning space into place. Subtitled *A Study of American Architecture and Civilization*, *Sticks and Stones* considers architecture as the relation between space and design, nature and culture. "Perhaps my most important contribution" to the study of architecture, Mumford later remarked, was that "I sought to relate individual structures to their urban site or their setting in the rural landscape."[22]

[22]See "Preface to the Dover Edition," p. ix, in Mumford, *Sticks and Stones: A Study of American Architecture and Civilization*, 2d ed. (New York: Dover, 1955; orig. pub. New York: Boni and Liveright, 1924). On the whole, Mumford judged American architecture harshly. He noted a general failure to create an authentic relationship between design and place. The overwhelming tendency is to remove design from setting. American architecture reflects the nonfunctional and irrelevant demands of those in pursuit of social status. Consequently, the direction has been toward even less satisfactory building. At least during the "classical" period of the early republic, beset as it was by canons of overly "abstract" and formalist design, there was an attempt to relate the design of the house to its urban setting. This was the age of "Baroque" city planning, with the emphasis on broad avenues and geometric grids. Later architecture was imbued with the "mortuary air of archaeology" present in the monuments of the early "imperial age" and the most dreary, modern manifestations of a "manufactured environment."

Mumford presented colonial New England as a conspicuous exception to the American pattern—a genuine place.[23] By treating setting and design, architecture and town planning as a whole, the builders of the New England town created an authentic form based on the organic principle, building in accord with nature. The design of the 17th-century structure gives one "the feeling not of formal abstract design, but of growth: the house has developed as the family within it has prospered. . . . There have been additions: by a lean-to at one end the kitchen has achieved a separate existence; and these unpainted, weathered oaken masses pile up with a cumulative richness of effect."[24] Equally important is the layout of the village. Compactness assured access to the religious life in the town center, and irregularity permitted the functional division of land into home use and farm sites in a manner compatible with the natural features of the land (see Figure 1). This distinctive character was not accidental. It was a consequence of the particularities of regional geography and in part the result of the inheritance of medieval practices of architecture and town planning not shared by other colonies. The layout of the village, made possible by both the exercise of town-planning authority and a mutualist ethos, was also organic to its geographic conditions. Mumford's colleague in the RPAA, Benton MacKaye, suggested that this village form might best be understood in biological terms: a "community" was manifested in its cellular structure—a town common, or "nucleus," and a mass that emanates from the center in a manner similar to such symmetrical natural forms as the starfish. More than architecture or town planning, the society is characterized by "cooperative ownership and direction": "the just design, the

[23]Mumford's emphasis on the medieval origin of New England's architecture and society and the consequent divergent pattern of settlement and social organization is borne out by historians' work. See Kenneth Lockridge, *A New England Town: The First Hundred Years* (New York: Norton, 1970); Sumner Chilton Powell, *Puritan Village: The Formation of a New England Town* (Middletown, Conn.: Wesleyan University Press, 1963); and Michael Zuckerman *Peaceable Kingdoms: New England Towns in the Eighteenth Century* (New York: Knopf, 1970), which is closest, I think, to Mumford's assessment.

[24]Mumford, *Sticks and Stones*, p. 8.

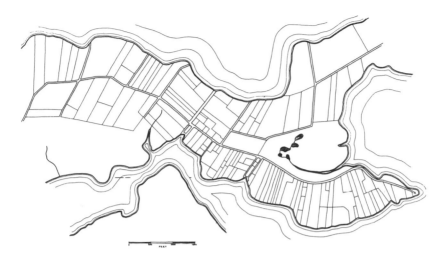

FIGURE 1. Plan of Salem, Massachusetts: 1670. While the New England town lacked many urban values, Mumford admired it for its compactness and irregularity; it reflected both a communitarian social tradition and the natural topography.

careful execution, the fine style that brings all the houses into harmony no matter how diverse the purposes they served . . . was the outcome of a common spirit, nourished by men who had divided the land fairly and who shared adversity and good fortune together." Or as MacKaye put it, the "structural symmetry" of the New England village "was equaled by its cultural symmetry."[25]

Mumford admired early New England as a *place* grounded in its geographic environment, in topos. It stood in contrast to the abstract organization of space of the typical American town: the geometric gridiron of streets pioneered in the plan for Philadelphia (see Figure 2). Because topographic influences were important there, New England suggested to Mumford the possibility of an authentic inhabiting of space.

Mumford did not suggest that the built environment determined the social life of a town; instead, he believed that the attainment of a genuine sense of place, grounded in a relation to nature, parallels—and encourages—community. In the early New England that

[25]Mumford, *Sticks and Stones*, p. 6; Benton MacKaye, *The New Exploration: A Philosophy of Regional Planning* (New York: Harcourt, Brace, 1928), p. 59.

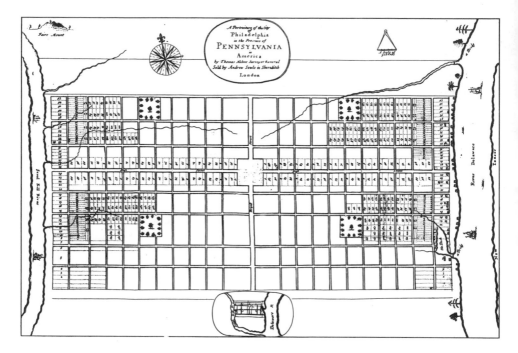

FIGURE 2. Plan of Philadelphia, Pennsylvania: 1683. The typical American plot map called for streets that had practical virtue but were imposed on the land without any regard for topography. This famous plan for Philadelphia helped establish the pattern.

Mumford admired, the techniques of building place—especially town planning—correspond to a culture of community: a commonality based on civic-mindedness and social cohesion. Thus, for Mumford, early New England meant: (1) *the sense of place* essential to establishing and maintaining a productive, equitable, and organic relation between the built and natural environments; and (2) the sense of the commons, defined by an acceptance of common destiny and by social relations based on dignity and mutual respect. These are the sociocultural characteristics necessary to civic identity in a vibrant regionalist democracy.

The importance of community in Mumford's work springs in part from his emphasis on place, on the natural environment, and on the relation of culture to nature. But his idea of community also reflects

Mumford's holistic insistence that social relations and the human-nature nexus must be related. He subscribed to the ecological principle of interdependence among species and between species and their common environment. Moreover, he drew a parallel between the natural and social ecologies. The mutualist social relations suggested by the idea of community help to define the regional interdependence of humanity and nature.

The regional folk culture and vernacular architecture of New England was adapted to the particularities of place. The village's social cohesion is evident in the use of town planning. It is from this community ethos that Transcendentalist individualism emerges. As Mumford saw it, Emerson embodied New England: "the preacher, the farmer, the scholar, the sturdy New England freeholder . . . the shrewd Yankee peddler or mechanic, were all encompassed by him" (*GD*, p. 95). Mumford defined Transcendentalism in relation to the culture: "what is vital in the American writers of the Golden Day grew out of a life which opened up to them every part of their social heritage" (*GD*, pp. 90, 93). Emerson's exercise of imaginative freedom was in line with social responsibility. His radical individualism was directed less at radical dissent against society and more toward defining a moral social order from within society. Emersonian "self-reliance," in other words, is compatible with the *community* of early New England.[26] Here *community* is defined as an ecology, characterized both by functional interdependence and by the attainment of

[26]This meant challenging earlier criticism of the Transcendentalists, especially that of Van Wyck Brooks, who had argued that Emerson's celebration of "self-reliance" was merely an expression of the ethos of the frontier, an irresponsible "individualism" of the soul that paralleled the acquisitive individualism of American capitalistic society. It is uncertain whether Mumford was aware of the criticism of Paul Elmer More; More's interpretation of Thoreau, in "Thoreau's Journals," *Shelburne Essays*, (New York: G. P. Putnam's Sons, 1910), also diverges from Mumford's. More saw Thoreau as an idealist, an interpretation later adopted by Marxist literary critics. While More celebrated this Neoplatonism, Marxist critics would despise it.

subjectivities that make possible mutual relations within the social and natural worlds.[27]

The fusion of the practical and the imaginative is another important measure of community. Mumford saw this by way of Emerson's "affirmation of both . . . science and myth." Emerson was the poet, the idealist and dreamer who nonetheless "did not make the mistake of disdaining the order and power that science had achieved within its proper department" (*GD*, p. 103). This is the message that Mumford hoped to convey: artistic creativity, "genius," is of little use if it does not make some real difference in the world. Later, in his study of Herman Melville (1929), Mumford discovered a model for the mediation of myth and science. *Moby Dick* stands as an "important synthesis" because it mirrors exactly what Mumford wanted to do in his cultural discourse: to create a multilayered representation of the human condition in which "every aspect of reality belongs." In particular, Melville's achievement confirmed Mumford's deepest hope that "science did not . . . destroy the myth-making [or imaginative] power of man, or reduce all of his inner striving to bleak impotence."[28] This speaks for Mumford's attempt to recontextualize modernity by holding science and imagination, logos and mythos, as parallel and essential discourses.

[27]Mumford's use of *community* parallels the investigations of primitive societies in the writings of American anthropologists Edward Sapir, Paul Radin, and Stanley Diamond. They all understood the value of the integrated cultures that manage to balance the inner world (the source of spirituality, creativity, and individuality) and the outer world of community (the source of social cohesion and mutuality). Maurice Stein, *Eclipse of Community: An Interpretation of American Studies* (Princeton: Princeton University Press, 1960), esp. pp. 230–250; Stanley Diamond, "In Search of the Primitive," in *Culture in History: Essays in Honor of Paul Radin* (Cambridge: Harvard University Press, 1960). For an anarchistic interpretation of the political implications of the transition from the primitive to the civilized, see Pierre Clastres, *Society against the State* (New York: Zone Books, 1989), esp. "Copernicus and the Savages," pp. 7–26, in which he argues for a typology of societies based on whether power is "coercive" or "non-coercive."

[28]Mumford, *Herman Melville* (New York: Literary Guild, 1929), pp. 170, 191.

The scientific studies of the region, social organization and town planning are Mumford's elaborations of "science," and the imaginative reconstruction of New England as a region is his appropriation of myth - the creation of a "usable past." Mumford's re-creation of New England in *The Golden Day* combines functionalist and phenomenological approaches to culture.[29] This approach addresses his central concern: the alienation of culture from nature. Mumford's imaginative resolution rests on the reconciliation of the natural region with the *social world*. For his sense of the organic was not a call for nature as wilderness but a plea "for joining the elements of Nature and Culture."[30]

Mumford's strength as a thinker lay in his holistic approach. This permitted him to cultivate a common ground for the social sciences and humanities, the aesthetic and practical worlds. This experiment was less a matter of theoretical certainty than of pragmatic aspiration. And that Mumford valued most the practice (not the theory) of reconciliation was indicated in his work with the Regional Planning Association of America.

When he turned to Whitman in *The Golden Day*, he was heartening himself for this effort. Mumford realized that Whitman best represented his vision and that with Whitman the golden day reached

[29]A project taken up by Paul Radin in *The World of Primitive Man* (New York: Grove Press, 1960): Radin describes how the technical and practical capacities of primitive cultures are confirmed and augmented by (though never made subordinate to) the magical and ritual practices. While Mumford's "premodern" world was not a primitive one, he shared Radin's understanding of the value of integration of "myth" and "science."

[30]Mumford's sense of the organic, his embrace of nature, may be compared to and contrasted with Sherman Paul's idea of a green tradition in American letters; see especially *Repossessing and Renewing: Essays in The Green American Tradition* (Baton Rouge: LSU Press, 1976). In this respect Mumford stands with Charles Olson. On Olson see Sherman Paul, *Hewing to Experience* (Iowa City: University of Iowa Press, 1989), pp. 215–225, esp. p. 222.

"high noon." He regarded Whitman highly because he took the Transcendentalist spirit and went into the world, discovering "mystical beauty" in all its dimensions (GD, p. 122). He consequently "absorbed so much of the America about him, that he is more than a single writer: he is almost a literature" (GD, p. 122). Whitman, claimed Mumford, was more inclusive than Emerson and Thoreau, and this inclusiveness put him in a better position to shape the culture, to engage it in a dialectic. Whitman absorbed the great "forces" of the modern world: "discovery [or science], industrialism and democracy" (GD, p. 130). But "neither political democracy nor industrial progress was for [Whitman] anything but a prelude to the third stage, rising out of the two previous ones, and creating a `native expression spirit'" (GD, p. 133). This "third stage" would be built from the "the possibilities of American life" (GD, p. 130). Whitman, more than Emerson or Thoreau, recognized the potential of this "third stage" to reconcile the power unleashed by industrial productivity with the practice of democracy. Such potential to renew America could find its realization in the simultaneous development of the sacred. For Whitman, Mumford noted, poetry took on "the office of sacred literature" (GD, p. 131): "American poetry was to do in our day what the Vedas, the Nackas, the Talmud, the Old Testament, the Gospel, Plato's works had done for their time: it was to crystallize our most precious experience and in turn to modify, by that act of crystallization, the daily routine" (GD, pp. 131–132). And in *Democratic Vistas* Whitman voiced Mumford's hope that "a cluster of mighty poets, artists, teachers" could become "national expressers, comprehending and effusing for the men and women of the States, what is universal, native, common to all."[31]

The "precious experience" that Mumford hoped to crystalize addresses the crux of the spiritual crisis of modernity, the alienation of culture from nature. In Whitman he rightly saw a poet as seer of the whole: Whitman's holism extended from the polis to the wilds; his

---

[31]Walt Whitman, *Democratic Vistas*, in Mark Van Doren, ed., *The Portable Walt Whitman* (New York: Penguin, 1977), p. 323.

inclusiveness spoke for Mumford's vision of nature in relation to culture. Attracted to the organicist strain in Transcendentalist thought, Mumford turned his vision toward polity and foresaw a new reciprocal relation of culture to nature: a regional democracy.

CHAPTER 4

Organicism
and Planning

By the time he had completed *Sticks and Stones* and *The Golden Day*,
Mumford had arrived at several conclusions that helped prepare him
for his work on regional planning. He was aware that reestablishing
an organic relation to nature is a holistic concern encompassing the
entire culture; that cultural criticism therefore requires a practical
outlet; that such a practical expression must take account of the
"forces" that have created modernity (most importantly, science and
technology); that biological and geographic sciences share common
concerns with organic principles of design; and that a technique of
regional planning incorporating an organicist understanding of art,
science, and life may be employed to redress the spatial imbalances
between built and natural environments. Consequently, Mumford envisioned a broad application for planning, grounded in the "underlying geographic and economic realities" of a region—a kind of
planning appropriate to the creation of "genuine communities" and
to achieving a balance between human activities and the natural world
(*C of C*, p. 363).

For Mumford, organicism was both content and method, an expression of the affirmation of nature that reflected his dual concern with the regional ecosystem and the philosophical principle of vitalism. It was also a way to hold oppositions, such as nature and culture, in relation to one another. Following Alfred Whitehead, we see that from an organicist perspective

> the notion of life implies a certain absoluteness of self-enjoyment. This must mean a certain immediate individuality, which is a complex process of appropriating into a unity of existence the many data presented as relevant by the physical processes of nature. Life implies the absolute, individual self-enjoyment arising out of this process of appropriation. I have . . . used the word *prehension* to express this process of appropriation. Also I have termed each individual act of immediate self-enjoyment as an *occasion of experience*. I hold that these unities of existence, these occasions of experience, are the really real things which in their collective unity compose the evolving universe, ever plunging into the creative advance.[1]

The "unity of existence" is fundamental to organicism: it is a holistic conception that uncovers the "really real" phenomena that constitute the universe. But to uncover this fundamental unity is not a process of finding essences or abstractions or ideals that stand above the plane of physical existence. Unity is embedded in continual "transition" of the universe. If there is an "essence of the universe," it is that "it passes into a future" (*M of T*, p. 152). The overall unity is therefore a "process" of creativity, a process made apparent in "prehension": prehension is made possible by "individual act[s] of immediate self-enjoyment." The capacity to "promote" the "fusion" of "data," or the material facts, with the hidden "potentialities" of life is a creative capacity grounded in "occasion[s] of experience."

For Mumford, this organicist idea of unity in process made necessary the work of combating the fragmentation of life and the specialization of knowledge. To begin this task, he developed his own

---

[1]Alfred Whitehead, *Modes of Thought* (New York: Free Press, 1968; orig. pub. New York: Macmillan, 1939), pp. 150–151; subsequently referred to as *M of T*.

organicist method of "fusion." His work was to advance ("promote") creative responses that "appropriated" the many data that experience presents and "fuse" them into an underlying unity, or "form." Form is essential, Whitehead reminded us, since "all actuality involves the realization of form" (*M of T*, p. 90). Mumford's method was to find expressions of form that might work (be functional) within the "occasion[s] of [modern] experience": forms—urban and regional plans and designs—that could provide a method of fusing life and art, aesthetic value and technological efficiency, logos and mythos. Consequently, he looked for forms that were functionally honest at the same time that they encompassed "imaginative design."

The planned, regional city is one such form, an expression of "the trend towards order which is the overwhelming deliverance of experience" an order that draws from the "boundless wealth of creative potentiality," a workable "alternative" to the actually existing built environment (*M of T*, pp. 88, 152). It is an alternative that begins with the possibilities of the machine (modern "technics") and proceeds by "complicating the technical in order to make it more organic."[2] The "complication" is a form of "prehension," the appropriation of different "data," or aspects of knowledge, into an organic whole that replicates ecological interrelatedness. In keeping with Whitehead's understanding that categories of knowledge are really "types of existence" (*M of T*, p. 95), Mumford saw that the processes that must be accounted for are biological and geographical—and also cultural and technological. The concern that underlay Mumford's search for workable alternatives was the "unity of existence": the possibility of finding ways to hold in relation to one another the fragmented worlds of "knowledge."

Mumford believed that an aesthetics of place could speak to modern technological and social conditions. The organicist concept of life as an imaginative and active *process* of creating (new) forms shaped Mumford's response to technological modernization, and this idea of process that bridges the technical and artistic worlds helped advance his phenomenology of place. The sense of place, Mumford

---

[2]Mumford, *Technics and Civilization* (New York: Harcourt, Brace, 1934), p. 376; hereafter referred to as *TC*.

insisted, has economic and cultural implications for the reshaping of the built environment. This is to say, as always in his work, that the aesthetic shapes the practical and technological and has moral and social implications. Like the writers of the golden day he admired, Mumford embraced the *possibilities* (if not always the *actualities*) of science and technology. He hoped that the very technological forces unleashed by the industrial revolution could be made compatible with organicist principles of design. For Mumford, the garden city— a modern city built with new technologies and sited in creative relation to the natural world—was a concrete expression of these hopes.

Thus genuine garden cities call forth the sense of place. If we can reestablish the importance of place to our sense of ourselves and to our understanding of life, we can conceive and invent new ways to utilize modern technology. Mumford thought of the region as the locus for the reconciliation of ecological design and technological modernization. He insisted that an entire region in "all its sites and resources, from forest to city, from highland to water level, may be soundly developed . . . so that population will be distributed so as to utilize, rather than to nullify or destroy, its natural advantages."[3]

In a 1928 essay for the *New Republic* entitled "Toward a Rational Modernism," Mumford called for the application of modernism in art to organicist ends. The "modern style," he said, can be "briefly defined: it is the emphasis of function and structure . . . simplicity and directness."[4] Modernism is related to the modernization of industrial production, for its "essential canon" is "that of respecting the function and acknowledging the structure" of machine-produced goods. Mumford appreciated this functional honesty as he explained in a 1921 *New Republic* article, the conventional attitude toward style is nothing more than "a pleasing superfluity that can be added to or

[3]Mumford, "Regions—To Live In," *Survey Graphic* 7 (May 1925): 151-152; reprinted in Carl Sussman, ed., *Planning the Fourth Migration*, (Cambridge: MIT Press, 1976), p. 90.

[4]Mumford, "Toward a Rational Modernism," *New Republic*, 25 April 1928, pp. 297–298.

withdrawn from a work of utilitarian art at will." He wanted a frank admission that "'style' is fundamentally an outcome of a way of living."[5] Modern design assails those dualities of life and art, the real and the ideal, the practical and the aesthetic, that made possible the creation of a degraded industrial sphere. For by separating the spiritual (aesthetic) from the practical (functional), 19th-century aesthetics had made both impossible. Like Frank Lloyd Wright (whom he cited) and Veblen, Mumford wanted to acknowledge "the modern material facts," chief among them the machine. Art must recognize this reality, and Mumford considers modernism a movement that promised to do so. Modern aesthetics permits us "to have our feet firmly planted in our own age." Thus Mumford understood that modernism, as an aesthetic and cultural movement, arose from the conditions established by 20th-century methods of production. In short, modernism could not, should not, be negated; it represented an understandable, and in some ways vital, response to the existing culture.

Yet modern design showed every sign of having "forgotten about man." In its enthusiastic embrace of a machine aesthetic, modernism had the potential (which was fully realized, especially in the later work of proponents such as Walter Gropius) of disregarding men and women in real communities. Mumford acknowledged this: "we must look for a finer relation between imaginative design and a whole range of biological, psychological, and sociological knowledge."[6] These (scientific) instruments of knowing would be useful inasmuch as they provide an understanding of the particularities of a "humane life" that should inform not only a modernist aesthetic but the use of planning as a technique to reconcile modern technologies with the natural world. Modernist architecture, in short, must not be built in isolation from either the natural environment or the creative imagination. Like Louis Sullivan, Mumford thought "that the origin of [an authentic] style is [not] outside, but within ourselves, and the man who has not the impulse within him will not have the style."[7] Cre-

[5]Mumford, "Machinery and the Modern Style," *New Republic*, 3 August 1921, p. 263.

[6]Mumford, "Rational Modernism," p. 298.

[7]Sullivan quoted in Sherman Paul, *Louis Sullivan, An Architect of American Thought* (Englewood Cliffs, N.J.: Prentice-Hall, 1962), pp. 40–41.

ativity is organic to self and nature; it must grow from within. Such creativity is the sine qua non of ecological design.

Though, much to his later embarrassment, he dismissed the Chicago School in his original edition of *Sticks and Stones*,[8] Mumford would undoubtedly have agreed with this description of Sullivan's principles of design:

> The identity of a building was not to be imposed but discovered and expressed. Form follows function was a rule of discovery and expression, and its implications would have been more clearly recognized if it had been stated: function creates form—"life seeks and takes on its forms in an accord perfectly responsive to its needs."[9]

In the modern industrialized world, where the perfection of *technical reason* means the subjugation of nature and regimentation of society, there is vital "need" to create a culture capable of extending and nurturing life in all its forms. This requires public deliberation on the ends to which modern technology is applied. To this end, Mumford sought to cultivate a sense of place necessary to community and to bring the built environment into balance with the natural region. He pointed out that "modernism may mean using a complicated engineering technology to produce cheap skyscraper apartments with elevators; or it may mean using comprehensive city-planning [techniques], as Mr. Henry Wright has done [at Sunnyside, a garden-city-like development], to avoid the necessity for skyscrapers and elevators."[10] In each case design reflects the modern capabilities for productivity: both take advantage of modern materials, large-scale production, and sophisticated planning techniques. But the results differ greatly. This distinction in form between urban megastructures and decentralized urban centers reflects a difference in the needs considered by the designers.

Inevitably, therefore, Mumford was drawn from aesthetics to tech-

[8]See "Preface to the Dover Edition," in *Sticks and Stones: A Study of American Architecture and Civilization*, 2d ed. (New York: Dover, 1955, orig. pub. New York: Boni and Liveright, 1924).

[9]Paul, *Louis Sullivan*, pp. 67–68.

[10]Mumford, "Rational Modernism," p. 298.

nology, from the possibilities of organic architecture to the potential of new technologies and techniques. This "neotechnics" might reverse the harm done to the natural environment by industrialization. Electric power (especially hydrogenerated electric power), the automobile, and the truck, as well as urban planning, would transform the built environment. These advanced technologies had the potential to overcome the environmentally destructive effects of the previously innovative technologies of the 19th century: the coal-powered steam engine, the railroad, and the industrial city. In the new technological developments, there was an incipient movement toward "industrial decentralization" so profound as to practically constitute an "industrial counter revolution."[11] It was counter to the spatial centralization of production and living, the organizing principle of the first industrial revolution—and to the principles of mechanization that had created it.

The origins of the industrial revolution and the prospects for a counterrevolution are the issues Mumford took up in 1934 in *Technics and Civilization* (by "technics," Mumford meant both technologies and the techniques that create them). The triumph of industrialization tipped the spatial balance toward the built environment and private space: cities ran roughshod over the natural region, and within the metropolis the provision for and use of public space declined, a reflection of the ascent of the private sphere (privacy, private "homes," and private economic power). Industrialism ushered in an era in which mining became "the dominant mode of exploitation" and "the pattern for subordinate forms of industry" (*TC*, p. 158). Rather than advancing the cause of the species, "paleotechnics" fostered an era of "barbarism" (*TC*, p. 154). The ongoing development of what Mumford called "neotechnics" represented his hope that the barbarism of the machine age would be transformed. And his reading of the past was intended to support his contention that to advance beyond the chaos of industrialization requires not a retreat to craft production but "the assimilation of the machine" to organic design. Mumford's central argument was simple: a new scientific technics is necessary, indeed essential, to make the machine

[11]Mumford, "Regions—To Live In," pp. 91–92.

compatible with life. When this is accomplished, "neotechnics" will pass into a true "biotechnics."

❖

Most important in this aspect of Mumford's work is his realization that the future of the human species depends on reconfiguring and recontextualizing the machine. Realizing that it is impossible to deny its moral impact, he argued that the machine must be humanized and naturalized. It is necessary, therefore, to reconceive the machine as an extension of living organisms. "The conclusion," Mumford reported,

> is obvious: we cannot intelligently accept the practical benefits of the machine without accepting its moral imperatives and its esthetic forms. Otherwise both ourselves and our society will be the victims of a shattering disunity, and one set of purposes, that which created the order of the machine, will be constantly at war with trivial and inferior personal impulses. . . . The real social distinction of modern technics . . . is that it tends to eliminate social distinctions. Its immediate goal is effective work. Its means are standardization: the emphasis of the generic and the typical. . . . Its ultimate aim is leisure—that is, the release of other organic capacities. (*TC*, pp. 355–356)

One task, then, is to set out the aesthetic and moral implications of mechanization: to bring the function of a machine economy to bear on the social and aesthetic forms of culture. The other depends on realizing that the machine itself must be recontextualized; it cannot be left (as it is generally understood) as an "abstraction" external to life with the power to reorder humanity and nature. Accordingly, Mumford developed multiple contextual understandings of the machine: social, aesthetic, and scientific. But science is the key to Mumford's project, not only because it had developed to the point that it could shape technological innovation (in the early 20th century the practical inventor had been replaced by the scientific researcher), but because science had undergone what Mumford thought was a revolutionary transformation: the focus on "the cosmic, the inorganic, the 'mechanical' . . . [had given way] to [concern with] every phase of human experience and every manifestation of life" (*TC*, p. 217). Here Mumford referred to the rise of classical sociology, the development of ecological science, and the systematic application of science to technological innovation and social concerns. Following Geddes, Mumford considered these conceptual and organizational changes to be a revolution in thought, one *likely to be* compatible with the organicist perspective. Given this reading, it is not surprising that Mumford envisioned an intellectual transformation so profound that it would overcome the mechanistic legacy of 19th-century science and technology.

Mumford redefined technics as a creative response to the forces of nature. He pointed out that the extension of science to the study of the geographic region allowed us to see that the rise of "paleotechnics" itself was in part a reasonable response to environmental conditions: the pre-industrial economy had depended on "eotechnics," power from water and wind. But the "dependence on strong steady winds and upon the regular flow of water limited the spread and universalization of this economy" (*TC*, p. 142). Wood came to be relied on for fuel and power, but its availability inevitably dwindled. The environmental limits to the *sustainability* of the old technological system necessitated the invention of a new source of power. This turned out to be the steam engine, the basis of the industrial revolution. But Mumford was careful not to reduce culture to the function of its machines. The rise of paleotechnic (steam-powered) industrialization was more than anything a change of mind: in its very conception paleotechnics rested on a new and disturbing worldview. For during this era Western culture abandoned organicism, substituting the new "realities" of mechanical power and economic growth for the geographic and biological imperatives of natural forms.

Technics and Civilization addresses this legacy of the machine by reconceiving it in terms of organic principles of function and design. The emphasis is on the present conditions and realities. Mumford responded to *his* time: he realized that the conditions *and expectations* created by the industrial revolution—the perfection of mass production, the capacity to support larger populations, the rising standards of living, the demand for a better life (manifest, for example, in the suburban exodus)—would have to be addressed. At the same time, recognizing the existence of natural limits to endless economic growth, Mumford asserted the need to conserve resources and work within the limitations and possibilities of the natural region. And both modern economics and natural limits and values could be addressed in terms of an organicist technics. He did not, therefore, accept the idea that ecological sustainability and economic development were mutually exclusive choices. The technological and cultural conditions of his time suggested to him that as a civilization we were (then) far from having to make such choices. There were other solutions.

But any such solutions would require the dissolution of the "mechanical ideology" that had shaped the machine since the industrial revolution: "The Western European conceived of the machine because he wanted regularity, order, certainty, because he wished to reduce the movement of his fellows as well as the behavior of the environment to a more definite, calculable basis" (*TC*, pp. 864–865). This ideology is manifest in the design and function of the mechanical contrivances of the paleotechnic era. The mechanization of warfare, the attempt to overwhelm nature, the dehumanization of social relations in mass society—all are reflections of the machine ideology, an "unquestioned faith" that has "left to the untutored egoisms of mankind the control of gigantic powers and engines [that] technics has conjured into existence" (*TC*, p. 366).

To counter this requires the "assimilation of the machine" to organic principles, a process Mumford (mistakenly) announced as already well under way: "we have now reached a point in the development of technology itself where the organic has began to dominate the machine" (*TC*, p. 367). The crucial transformation involves the replacement of the "neutral valueless world of science" (*TC*, p. 367) with a new vitalist and holistic science: this represents the anticipated shift in science from "the world . . . conceived as a series of independent systems" to a "single system" in which the "primary data are social and vital" (*TC*, pp. 369–370). These data point to an organicist prospect that is "life-furthering," that replaces the conception of humans and their machines operating in a "blind and meaningless universe" with the idea that human endeavor may become "a partnership in mutual aid" with the earth (*TC*, p. 370). The most important discovery of the new vitalistic science is the study of the "ecological balance of the region": a new kind of scientific study, confirmed in Aldo Leopold's *Sand County Almanac*, which finds expression in the understanding that "the region . . . [has] some of the characteristics of the individual organism: like the organism, it [has] various methods of meeting maladjustment and maintaining its balance." By contrast, mechanistic science turned the region "into a *specialized machine* for producing a single kind of goods—wheat, trees, coal . . . [and thereby we came] to forget its many-sided potential as a habitat for organic life [and] . . . finally . . . to unsettle and make precarious the single economic function that seemed so important" (*TC*, pp. 256–257, emphasis added).

Holism—Whitehead's "unity of existence"—is the remedy for the fragmented and mechanistic worldview of conventional (19th- and 20th-century) science. It enables us to see that "form, pattern, configuration, organism, historical filiation, ecological relationship are concepts that work up and down the ladder of the sciences" and therefore that "esthetic structure" and "social relations" are inherent in scientific knowledge (*TC*, p. 371). The idea that science has social and aesthetic implications recalls Dewey's pragmatism and explains the importance of the aesthetic and moral points of view in Mumford's work. It sustains his hope for a synthesis of logos and mythos. (For what is myth if not an "esthetic structure"?) And this is precisely why so much of *Technics and Civilization* concerns the aesthetics of modernist design; Mumford was in search of openings for organic and holistic patterns that were indicative of what he foresaw: a paradigmatic shift of science and technics. Thus he was heartened by modernist functionalism, with its honest representation of the machine's operation. And he argued that modernist design may be made to further suit organic principles by seeing it in an environmental context. Good design not only reflects the techniques of its production (functionalism) but considers the social and natural settings.

❖

As important as aesthetics and science were to Mumford's measure of the organic, he was well aware that "one knows life . . . only as
one is conscious of human society." (*TC*, p. 370). And Mumford understood very well the social consequences of industrialization. The
exploitation of workers everywhere attended the paleotechnic order.
The increased use of machinery "reduced" the worker "to the function of a cog," a person "bound to the machine" by means of "starva-
tion, ignorance and fear" (*TC*, p. 173). Already beset in the
pre-industrial era by increasing specialization and the division of
labor, the worker became in the era of industrialization a victim of
the factory, "the most remarkable piece of regimentation . . . that the
last thousand years have seen" (*TC*, p. 174).[12] Thus even as "industry became more advanced from a mechanical point of view it . . .
became more backward from a human standpoint" (*TC*, p. 146). This
relates to capitalism itself, which made "money-getting" a "specialized form of activity" and profit the "main economic objective" (*TC*,
p. 373):

> what are called gains in capitalist economics often turn out . . . to
> be losses; while the real gains, the gains upon which all activities
> of life, civilization, and culture ultimately depend were either
> counted as losses, or they were ignored, because they remained
> outside the commercial scheme of accountancy. (*TC*, p. 375)

This came about because the new "realities were money, prices,
capital, shares: the environment itself, like most of human existence,
was treated as an *abstraction*" (*TC*, p. 168, emphasis added). As this
analysis implies, there was a moral failure of the capitalist order—a

---

[12]Recent histories, such as David Montgomery, *The Fall of the House of
Labor* (Cambridge: Cambridge University Press, 1987), p. 13, have shown
that "skilled craftsmen exercised an impressive degree of collective control
over specific productive tasks in which they were engaged," a power over
the work process that was *not* inherited. Instead, "they exercised this control because they fought for it." This autonomy was gradually eroded in the
period of mass production of the early 20th century—the very period during which Mumford was writing.

corruption of the real in favor of the abstract, a capacity to destroy life in favor of the regularities of a mechanistic order. In sum, there was a failure to make the organic principle a cardinal measure of human activity.

To overcome this, Mumford argued that the rehabilitation of the intellectual, aesthetic, and moral principles of the Western worldview—a process Mumford identified with the transformation from paleo- to neotechnics—should find an outlet in a different pattern of social relations. In the last chapter of *Technics and Civilization*, Mumford called for the exercise of public power in a way that would fundamentally change capitalism. He argued that the gains in technological design can be realized only in the socialization of much that was then (and is now) left to the private sphere and the incentives of individual gain. Specifically, he called for "making a socialized monopoly of all raw materials" necessary to energy production; for "common ownership of the means of converting energy"; for land use controls to retain agricultural land and encourage specific kinds of farming; for planning to make "maximum utilization of those regions in which kinetic energy in the form of sun, wind, and running water is abundantly available"; for a "genuine rationalization of industry" that requires "the reduction of trivial and degrading forms of work"; for "the elimination of products that have no real social use"; for a "conscious economic regionalism" predicated on finding regional balances that "combat the evil of over-specialization"; and, finally, for a system of production that aims at accepting the limits of human wants and focusing on satisfying basic needs rather than perfecting a system of commodification designed for an ever expanding increase in levels and expectations of consumption (*TC*, pp. 380–382, 385, 388–389, 392–397).

Mumford grounded these powers in the "community" or public interest, reminding us that "the energy, the technical knowledge, the social heritage of a community belongs equally to every member of it, since in the large the individual contributions and differences are completely insignificant" (*TC*, p. 403). And he envisioned the possibility of having the state embody the community's interest. He even went so far as to call for the nationalization of banking, the reorganization of trade unions, "the organization of industry within the political framework of cooperating states," and the creation of consumers' organizations. It is unfair, therefore, to say that Mumford lacked political awareness. But he did tend to make politics subsidiary: necessary political change flows from the reconceptualization of knowledge. If we are now free to "work out the details of a new political and social order," it is *by reason of the knowledge* that is already at our command" (*TC*, p. 417, emphasis added). Although Mumford knew that society would have to be changed in order to realize the promise of an organicist science and technology, he understood these institutional changes as secondary and derivative. They would occur "by reason of knowledge."

In *Technics and Civilization* it is the imaginative and technical failures that are the crucial ones; the social and political failures of the capitalist order are left as implicit. The negative social consequences of machine production arise from and flow through the design and conception of the new technologies: the factory is, first and foremost, a failure of design and conception (idea). The discordant and unjust social and political consequences of the design are considered derivative and secondary. Furthermore, Mumford represents "neotechnics" as the development of biological and ecological science; largely assuming that these sciences would be organicist in their orientation and that they would transform the social order.[13] How else can we explain the authoritative tone of *Technics and Civilization*, the certainty with which Mumford predicted the course of technical and scientific development toward "biocentric" civilization? He made it seem as if the organicist and vitalist principles he championed were implicit within neotechnics. He seemed to forget his realization in *The Golden Day* that science itself had to be recontextualized, its Cartesian worldview challenged and supplanted.

Surely Mumford was right to anticipate a new paradigm for understanding science and technology, and he was right to understand that this is essential for reshaping the human condition. But to argue, as I think Mumford did at times, that imagining a systematic

[13]For a discussion of the impact of new paradigms of scientific thinking, including chaos theory, on ecological science, see Donald Worster, *The Wealth of Nature* (Oxford: Oxford University Press, 1993), pp. 156–170.

synthesis of science and organic vitalism in itself will have such an effect is folly. Rather, I agree with Whitehead: just as there are no definite "laws of change," there is no underlying *system of knowledge* from which change will spring. While Mumford would agree with Whitehead that any holistic achievement is always shifting and contingent, that our search for "new types of existence" (*M of T*, p. 95) is possible only as a pragmatic exploration, Mumford's early theory of science and neotechnics often belies these truths.

The difficulty in fully endorsing Mumford's views must not obscure his important contributions: he understood how technological innovation can address environmental ills of existing technologies, he saw the linkage between the form and function of technology and ideas, and he appreciated the importance of bridging the artificial divisions between branches of knowledge. The difficulty is that Mumford was not explicitly attentive, especially in *Technics and Civilization*, to the political dynamics of his ideas. While it is important, indeed crucial, to see the significance of Mumford's hope for neotechnics, a prospect he rightly believed required the recontextualizing of science, we have to admit that Mumford does not adequately explore the political and economic contexts for the promise of neotechnics. Consequently, he ascribed greater significance and hopefulness to the rise of the new scientific elites than was warranted.[14]

[14]In *Beloved Community* (Chapel Hill: University of North Carolina Press, 1990), pp. 284–285, Casey Blake argues that Mumford celebrated the "emerging power of professionals and managers within industry and government . . . [while] his historical investigations traced the ascendancy of such groups back to the acquisitive and predatory impulses of a militaristic early capitalism, which had crushed early craft traditions and local cultures. His neotechnic prophecy looked to the very same cultural forces to create a more humane industrialism. . . . [And his endorsement] of new forms of industrial planning and human relations organization [introduced by Frederick Taylor and Elton Mayo] reflected [his] profound hostility to politics." Blake is right about Mumford's assumptions about elites, but Blake may also underappreciate the organicist context that Mumford attached to neotechnics.

Mumford's discourse on science and planning is essential to his understanding of the social world. What he failed to demonstrate in *Technics and Civilization* is that the return to a "civic" consciousness requires more than working out a neotechnic economy or even an organicist experience of knowing. It also demands a discourse capable of shaping political and economic contexts that can support these changes. In his early work Mumford linked the process of creating place to the recovery of what Christopher Lasch calls "culture" as against "civilization": "the old organic solidarity of the preindustrial village" against "modern individualism and anomie." In this sense Mumford shared the common concern of many early-20th-century social theorists with reviving the "old solidarity" of premodern society on "a new basis" suitable to modern technologies.[15] But again, the political implications of this organic social vision are far from clear. At times, Mumford argued that this "community" is the outcome of the aesthetic-technical process of planning; "community" invokes a backward vision of premodern social life to describe a *place* that is "a permanent seat of life and culture," a "new environment on the human scale" that may become a "local center of culture." A genuine regional garden city is a planned community, one that manages to create a new *form* from the rudiments of the aesthetic and technological synthesis that Mumford envisioned. Having laid this basis, the planned community generates the conditions ideal for the creation of "counter-institutions": cooperatives, "little theaters" and new experimental schools—in short, an alternative society conducive to the "community spirit" growing inside the shell of the planned community.[16]

In his later work, Mumford increasingly emphasized the importance of a public sphere—a public space—that is democratic and capable of shaping a civic-minded social world. If the vision of ecological regionalism is to have implications for American society, it must find a political *form* capable of reshaping public discourse. The

[15]Christopher Lasch, *The True and Only Heaven: Progress and Its Critics* (New York: Norton, 1991), pp. 136, 139.
[16]Mumford, "Regions—To Live In," pp. 91, 93.

civic institutions of an alternative society cannot be planned in advance; they must emerge alongside planning, thereby revitalizing, regions as part of the larger American republic. Mumford's understanding of the importance of politics to ecological planning grew out of his experience of the RPAA's struggle to bring garden city regional planning to fruition.

PART II

Undertaking A Vision

Planning has a built-in contradiction. It is, as Gunnar Myrdal said, "an exercise in a non-deterministic conception of history" permitting a freedom of choice which may turn the course of future development. On the other hand, planning even on its simplest level, tends to be inimical to future freedom of choice, since the plan requires the fixing of boundaries, the channeling of actions, and the fixing of goals. The contradiction cannot be resolved by not planning, as a decision not to plan is also a plan.

Percival Goodman
The Double E

CHAPTER 5

"Regions—to Live In"

Even before he completed *Sticks and Stones* and *The Golden Day,*
Mumford's abiding interest in regionalism and his underlying hope
that science and technology could be redirected toward ecological
ends thrust him into the public sphere. In March 1923 he met New
York architect Clarence Stein to talk about garden cities and regional
planning. Out of their discussion came a proposal for the formation
of an organization dedicated to finding a means to reconstruct the
"physical environment of city and country life." Stein became the
chief promoter and organizer of the Regional Planning Association
of America, but Mumford was its chief spokesman. In addition to
Mumford and Stein, the RPAA included Stein's architectural partner,
Henry Wright, architect Frederick Ackerman, economist and writer
Stuart Chase, and forester Benton MacKaye. These six men, with the
later addition of housing advocate Catherine Bauer, constituted what
Roy Lubove calls the group's "inner circle."[1] Eventually the RPAA
had more than thirty members, most of whom, like Stein, Wright,
and Ackerman, had been members of the Committee on Community

[1]See Roy Lubove, *Community Planning in the 1920's* (Pittsburgh: University of Pittsburgh Press, 1963), pp. 38–41.

Planning (CCP) of the American Institute of Architects. Under the leadership of Charles Harris Whitaker, editor of the *AIA Journal*, the CCP had been especially active during World War I, promoting social architecture and community planning; it saw the building of planned communities as a powerful implement for progressive social reform.

Not surprisingly, then, when Stein and Mumford met they shared one central interest: planned communities and garden cities. The garden city idea was originally conceived by Ebenezer Howard in a popular little book he entitled *Tomorrow, a Peaceful Path to Real Reform* (1898).[2] An Englishman whose desire for a new life had led him to homestead in Nebraska, Howard returned home with a fresh contribution to his country's urban and social problems. Knowing that English cities were badly overcrowded and in need of additional housing, he proposed that instead of building housing within the cities, as many reformers proposed, new garden cities be built in the countryside surrounding major urban areas—not as suburb or bedroom communities but as comprehensive cities enclosing industry and commerce and providing for civic functions. Each new city would be planned as a self-contained urban area capable of providing its inhabitants complete employment and shopping and cultural facilities. Howard envisioned ringing London with garden cities to relieve its overpopulation.

For Stein and many of his architectural associates, the garden city represented an opportunity to implement new practices in design of housing in relation to site: modernist architectural values could be realized through unity of design and the utilization of modern methods of mass production. For Mumford, this meant two things. First, that the potential for using some of the techniques of modern indus-

[2]Ebenezer Howard, *Garden Cities of To-morrow*, ed. by F. J. Osborn and intro. by L. Mumford (Cambridge: MIT Press, 1965); first pub as *Tomorrow, a Peaceful Path to Real Reform* (London: Sonnenschein, 1898). As indicated by the original title, Howard thought of the garden city as an alternative to a socialist revolution.

try—mass production, large-scale planning, a unified conception—could best be realized in a new setting such as that provided by a garden city; and since housing, like other material needs, is often in short supply, the application of large-scale production to the creation of housing could alleviate this problem. Second, by using modern architectural and planning techniques, the garden city could recontextualize modern industrial productivity. Rather than simply producing more of the same—bigger cities, taller buildings, faster means of transportation—the garden city could provide an entirely different urban context, one characterized by *balance*: a "functional balance" of industry and residence, commerce and civic structure—all encompassed within a spatial arrangement that restores a sense of human scale to urban life. And the garden city was the best hope for reviving the larger natural context of the city by helping to restore and renew the viability of the surrounding natural and cultural *region*. Mumford believed the garden city could be a "regional city": a new kind of modern city in creative relationship with the surrounding countryside. More than an aesthetic embellishment of urban civilization, the countryside would have a direct relationship to the city. Urban life would not cease to exist but would simply have a different context (actually a very old one). The natural world would be felt: the garden city would provide rural landscapes, agricultural products, and electric power; it would nurture architecture and literature as well as particular kinds of industry.

Thus while the garden city became an important part of the RPAA agenda, Mumford kept the group focused on the centrality of regionalism. When Stein proposed that the group be called "Garden City and Regional Planning Association," Mumford dropped "Garden City" from the name in order to emphasize regionalism as a "fundamental philosophy" guiding the group's exploration of the "technique of creating new environments."[3] Regionalism became the governing principle of the RPAA, and planning its central technique. The RPAA stood for planning as an instrument of change, a means to

[3]Stein, "Proposed Garden City and Regional Planning Association," unpublished MS, 7 March 1923; and Mumford to Stein, 14 March 1923, Clarence S. Stein Papers, Olin Library, Cornell University, Ithaca, New York.

reconfigure the built environment in a manner conducive to the creation of communities within balanced regions.

▦

Notwithstanding their different experiences and perspectives, the members of the RPAA felt comfortable enough with one anothers' points of view to respond to Mumford's call for a unified vision of regional planning. The result was a special issue of the *Survey Graphic* magazine published in May 1925. Taken collectively, the articles in this issue present a vision of regionalism and a practical program for urban and regional planning. Two essays by Mumford set the tone. Calling for a form of regional development compatible with a rising regional consciousness, Mumford saw the possibility of overcoming the estrangement of urban life from the rural world. Mumford's central theme, the four "migrations" of the American people, indicated his belief that an historical opportunity for regionalism and regional planning was at hand.

In "The Fourth Migration," drawing on the cultural analysis in *The Golden Day*, Mumford pictured 19th-century Americans as a band of "restless" pioneers who experienced an "unprecedented urge to take possession of the continent." In a very short time, men and women scattered to the ends of the continent; their achievement was expensive, for while they "planted farms," they also "skinned the land." This was the first of the four great migrations. As the first migration peaked in the mid-19th century, a second migration that drew both Americans and European immigrants to urban areas to work in factories was getting underway. Industrialization transformed the old commercial city (e.g., Philadelphia) and created a new industrial city (e.g., Pittsburgh), developed as a machine for production. Producing vast quantities of goods, the new industrial city became "solely . . . a place of work and business opportunity." The third migration began in the late 19th century when "productive effort came to take second place to financial direction." The consolidation of industries, the growth of banking and stock markets, the development of advertising, and the creation of a national market spurred the development of the large metropolis (e.g., New York and Chicago), whose rapid growth was at the expense of surrounding and peripheral regions. The ensuing spread of "metropolitan culture" deprived other regions of the opportunity of genuine cultural—and economic—development. A "technological revolution that has taken place during the last thirty years" produced another "tidal movement of population" outward, from the large cities into adjacent areas. This fourth migration gave Americans an opportunity "to remold themselves and their institutions"; at the same time the migration reflected the judgment of history regarding the fate of the metropolis. Since the "forces that created the great cities [have] made improvement within them hopeless," the new effort "to build up a more exhilarating kind of environment" must begin by building garden cities beyond the metropolis. Only by admitting that the "hope of the city lies outside itself" will we be able to establish the regional city: "a permanent seat of life and culture, urban in its advantages; permanently rural in its situation."

This regional city, however, is only part of the reorientation of social, cultural, and economic life "towards a higher type of civilization": a realignment that begins with conservation and preservation—the establishment of "forest culture as against forest mining . . . permanent [sustainable] agriculture instead of land skinning" and rural landscapes instead of suburban sprawl. Conservation suggests new economies—"specifically planned for the maximum of local subsistence . . . [with] communities based on natural economic and geographical considerations"—and the revitalization and diversification of rural and urban economies.[4]

Mumford envisioned the RPAA as the spearhead for the reconstruction of *both* urban and rural America. MacKaye's proposal to utilize a hiking trail through the Appalachian Mountains to spur a new rural economy served as a paradigm for rural development, just as Stein and Wright's work on housing and urban development contributed to the RPAA's hopes for realigning urban regions through planned decentralization. It was MacKaye's Appalachian Trail, though, that established the model for the RPAA: the linkage of regional devel-

[4]Mumford, "The Fourth Migration," and "Regions—To Live In," *Survey Graphic* 7 (May 1925): 130–133, 151–152; reprinted in Carl Sussman, ed., *Planning the Fourth Migration* (Cambridge: MIT Press, 1976), pp. 55–64, 89–93.

opment and planning to the recovery of regional consciousness. This idea lay at the heart of the RPAA's hope that the application of modern technology would be compatible with the geographic and cultural integrity of diverse regions. MacKaye envisioned rural populations' enjoying the benefits of industrial employment, perhaps on a part-time basis, while also benefiting from a rural location that allowed them to have gardens and follow other rural pursuits and to maintain a distinct cultural identity. Rural areas could bear industrialization without significant environmental damage because they would be home to a new form of industry, decentralized and planned rather than diffused and chaotic. New technologies and regional planning would make the difference.

The RPAA argued that this program for planned decentralization was compatible with the process of industrial diffusion that had already begun to affect American urban development. Critical to their line of reasoning was the group's enthusiasm for new technologies: automobiles, hydroelectric power, and electric transmission lines. Of these, none was more important to balanced regional development and planning than the electric transmission line. In the 1920s rural America was just emerging from relative isolation. Automobiles, limited to the prosperous agricultural regions, made possible increased contact with "town" (though roads were still rough). Radios had become available, but most families in rural areas did not have electric service. As Robert Bruere, editor of *Survey Graphic* and RPAA fellow traveler, explained in his article "Giant Power," the new capability to transmit electricity over long distances was the key to the development of rural regions. Bruere recounted the experience of the provincial government of Ontario in harnessing Niagara Falls and forming a public commission to distribute power throughout the province. A study authorized by Governor Gifford Pinchot of Pennsylvania, recommended a similar course of action for that state.[5]

[5]During the 1920s a debate raged over a plan for the construction of a power plant at Muscle Shoals, Alabama. The southern power companies were not interested in the plant, and conservatives resisted the principle of state involvement. Progressives countered with the argument made by Robert Bruere in "Giant Power —Region-Builder," *Survey Graphic* 7 (May 1925): 161–164, 188; reprinted in Sussman, *Planning the Fourth Migration*, pp. 111–120: Only public control will ensure fair distribution of power to remote areas.

The result of widely distributed electric power was apparent: diffusion of industry and population. As Mumford put it, instead of "being tethered to the railroad and its coal shipments, industry can move out of the railroad zone" into new regional industrial centers. The RPAA lavished similar praise on automobiles and electronic communication, which could bring the benefits of civilization into the most remote locations.[6]

The new technologies accelerated the dispersal of urban functions and led to the development of a new pattern of urbanization. In his *Survey Graphic* articles, Mumford welcomed this emerging pattern as an opportunity to create a balance between urban and rural areas by giving outlying areas a fair chance to attract industries and populations that had flocked to the overbuilt cities ever since the industrial revolution. But it was essential, he maintained, to direct the development of these industries into new regional cities, thereby avoiding aimless scattering. Because regional cities would be planned with definite limits on size and population, the dispersal of industries and people from major cities could become a structured decentralization. Once developed, a regional city would not be permitted to grow; additional population would be accommodated by constructing new cities. Since experiencing the physical boundaries of a community is important to a sense of place, those boundaries should be enhanced by geography: each new city would be surrounded by an agricultural and recreational zone called a greenbelt. But in addition to delineating boundaries (and avoiding the phenomenon now called urban sprawl), the greenbelt would provide each city with its own source of fresh fruit, vegetables, and milk. To balance the urban way of life by providing greater contact with rural life, regional cities would be designed to encourage mutuality and interdependence between farmers and urban residents.

Building planned regional cities as centers of new development would permit the industrialization of rural areas in a manner compatible with the character of the natural and cultural landscape. Rural areas of the United States would experience urbanization in an innovative form that preserved rather than destroyed regional geog-

[6]Mumford, "The Fourth Migration," p. 63.

raphy and culture. At the same time large metropolitan areas could use the garden city idea to decentralize, to unburden the central city of population pressures, and to preserve open land within proximity of the center.

❖

The most important single document the RPAA produced was the *Report of the New York State Commission of Housing and Regional Planning (CHRP Report)*. Authored by Henry Wright with assistance from the members of the RPAA's inner circle,[7] the *CHRP Report* is both an historical account of New York State's geo-economic development and an application of RPAA principles of regional development. It demonstrates the significance of the group's effort to be of direct influence in the ongoing process of urban redevelopment.

The *CHRP Report* emphasizes the environmental and technical factors in the development of New York State. After a period of early development during which natural resources such as fur and lumber were heavily exploited, an epoch of stability followed; in this era the technology and economy closely paralleled what Mumford called "eotechnics." According to the *CHRP Report* this early phase lasted until 1840 and "was characterized by almost complete noncentralization, with small, self-sufficing communities scattered throughout the State—each raising its own food and manufacturing the greater part of its own necessities" (*CHRP*, p. 155). In large part because of its self-sufficient economy based on agriculture and the harnessing of widespread waterpower (produced by small water mills), a regional culture and a decentralized, egalitarian society had

[7]State of New York, *Report of the New York State Commission of Housing and Regional Planning to Governor Alfred E. Smith* (Albany: J. B. Lyon, 1926), was made possible by Stein's relationship to the Democratic Party; after the re-election of Al Smith as governor of the State of New York in 1922, Stein was appointed chair of the commission; Carl Sussman, who reprinted the report in *Planning the Fourth Migration*, pp. 145–194, credited its authorship to Henry Wright, though he added that all the RPAAers had a hand in it. It is probably the single most representative piece the group produced. Subsequent references will be to the *CHRP* and will refer to Sussman's page numbers.

developed. But the coming of state-wide transportation systems, first canals and then railroads, brought economic specialization, centralization of industries, and an end to the first era. The *CHRP Report* divides the new period, which lasted from 1840 until 1920, into two equal "epochs." Overall, the trend during those eighty years was toward decreasing farm acreage as a percentage of all land use, agricultural specialization (fruit, dairy) as the West began to supply the grain market, concentration of the population in towns and cities accessible to the trade routes, increasing economic specialization, and, finally, large-scale urbanization. Of fundamental importance here is the *CHRP Report*'s explanation for these changes: "the growing importance of steam driven machinery both enabled and *dictated* this change in economic structure" (*CHRP*, p. 157; emphasis added). Steam depended on shipments of coal, which in turn depended on the railroad lines; these terminated in large cities connected to world markets by shipping.

By contrast, the *CHRP Report* predicts that the third epoch would be one of decentralization, presaged by the growing fourth migration of urban residents to suburban locations. Decentralization was a response both to the degradation of the urban environment and an opportunity to apply innovative technologies, especially in the areas of transportation, communication, and energy production. While urban concentration was the primary characteristic of the first two epochs of New York State's development, the RPAA argued that the social cost of urbanization and the economic burden imposed on business by spiraling land costs indicated that urban diffusion would be the most dramatic development of the new epoch: "we now appear to be on the threshold of a period in which a strongly marked new trend will be established" (*CHRP*, p. 176). The *CHRP Report* concludes that only "intelligent direction and coordination of future development" (*CHRP*, p. 177) would ensure a workable pattern of decentralization.

CHAPTER 6

Regional Planning as "Exploration"

A central tenet of the RPAA—that landscape can be restored as a "cultural resource," a compelling point of contact between nature and culture—was at the heart of the regional consciousness Mumford advocated. The recovery of landscape was a central component in the RPAA's attempt to reconceive and reorient the relationship between the built and natural environments. While landscape concerned the aesthetic sensibility that was at the core of regionalism, it was inseparable from planning for a reorientation of cities, a visionary planning that answered Mumford's early call for "utopias of reconstruction."[1] Yet even in its most visionary hopes, the RPAA guarded against making "plans that do not arise out of real situations . . . [which become] mere utopias of escape" (*C of C*, p. 390).

Drawing in part on a model created by his colleague Benton MacKaye, Mumford pushed the RPAA to link the issues of landscape

[1] In his earliest book, *The Story of Utopias* (New York: Boni and Liveright, 1922), Mumford distinguished between two kinds of utopian thinking: reconstructive and escapist.

and regional design with the possibility of shaping the structure of industrialized economies. Realizing that industrialization had yet to sweep the entire globe, Mumford hoped that planned regional development in the United States might establish a paradigm for accommodating large-scale industry and urbanization without overwhelming the existing natural and cultural landscapes. Developing countries would have a model to draw upon that would save them from the ills of Western-style industrialization and urbanization. In *The Culture of Cities,* Mumford remarked that "the real opportunity for urban and regional development lies in the fact that the existing [centralized, old industrial] pattern of economic life cannot remain stable" (*C of C*, p. 391). Having read Kropotkin's *Fields, Factories and Workshops*, Mumford understood modernization and industrial diffusion as in part, responses to increasing global competition. The unintended result (then as now) is overproduction resulting in a glut of manufactured goods. In response to the need to stimulate demand for products, corporate interests in the industrialized world turn to a powerful mix of colonialism abroad and advertising at home. This creates what Mumford characterized as a parasitic relationship between "over-urbanised communities" (the metropolis) and the "exploited regions" from which raw materials are drawn. By positing a new relation between production and consumption, regionalism would establish an alternative to the present configuration of the world economic system.[2]

Mumford recalled the medieval town, with its strong relationship to the surrounding countryside: "what gave the early medieval town a sound basis for health was the fact that, though surrounded by a wall, it was still part of the open country. . . . Nor were the towns, for centuries to come, wholly industrial [or commercial]: a good part of

[2]Mumford, "Regionalism and Irregionalism," *Sociological Review* 19 (October 1927): 277–288, esp. 286. Mumford identified the hidden principle of industrialization as an extreme form of geographic specialization that is antithetical to ecological balance; this system has undermined the ecological basis of life and the unity and equality of a healthy society.

the population had private gardens and practiced rural occupations" (*C of C*, p. 42). The relationship between country and city that Mumford admired is cultural *and* economic; it provides urban dwellers with a sense of identity in place that incorporated rather than excluded the surrounding countryside. Regional consciousness encourages the urban dweller to think about and interact with the natural world. This challenges classical economics, with its conception of the free market, a way of thinking and living that has moved us further and further from local sources: economically by forcing an artificial reliance on distant goods and markets and culturally through the internationalization of styles and tastes.

Regionalism, accordingly, is not only an aesthetic movement. A "balanced" region requires effective institutions that promote community interests as well as its own agricultural and energy resources. Mumford recalled that, in the first garden city built in England prior to World War I, the greenbelt was conceived as an "agricultural estate" of the city, providing dairy products, fruits, and vegetables and thereby establishing a direct functional/economic relationship between city and surrounding countryside.[3] He admired Ebenezer Howard because "his vision was bi-focal: he saw the countryside as well as the city . . . [and understood that] the problem of bettering life at both poles was a single one" (*C of C*, p. 395). Mumford recognized, furthermore, that the potential for decentralizing the production of energy is crucial to the prospects of genuine regional development: "the availability of water-power for producing energy . . . changes the potential distribution of modern industry throughout the planet" (*TC*, p. 222). Decentralized energy may become the

[3]See Ebenezer Howard, *Garden Cities of Tomorrow*, ed. by F. J. Osborn and intro. by L. Mumford (Cambridge: MIT Press, 1965); also Robert Beevers, *The Garden City Utopia* (New York: St. Martin's Press, 1988), pp. 44–47. In *The Last Landscape* (Garden City, N.Y.: Doubleday, 1968), pp. 135–162, William Whyte rightly criticizes garden city planners for conceiving of greenbelts largely as empty spaces, lacking a defined *geographic* importance except to delineate the boundaries of cities. Whyte admits that in Howard's original conception, the greenbelt played the positive function of providing open space for urban dwellers, but he fails to point out that Howard also conceived his greenbelts as the basis for sustaining a regional market for agriculture.

basis of local and regional economies. Mumford appropriately united the ideas of *natural landscape* and *economic productivity*. The balanced region combines both: it is a way of living that overcomes the increasing functional differentiation of space that has characterized the metropolis and led to the fragmentation of living/feeling from working/producing.

The regionalism of *The Culture of Cities* cannot be understood apart from the work of MacKaye. His ideas, outlined in the article "An Appalachian Trail: A Project in Regional Planning," published in the *AIA Journal* in 1921, should be considered the cornerstone of the RPAA's principles. In fact, it was MacKaye and Mumford who pushed the RPAA toward inclusive principles of regional planning, beyond the more limited scope of the architect-planners, whose initial interest centered on housing issues; moreover, MacKaye and Mumford were responsible for conceiving the special issue of the *Survey Graphic* that developed the regional basis for garden cities and urban planning.[4]

A professional forester, MacKaye had been employed by the U.S. Forest Service, beginning his career under the famous conservationist Gifford Pinchot. Like Pinchot, MacKaye had been trained at Harvard and was rooted in the "conservationist" ethic, what one historian called the "gospel of efficiency."[5] MacKaye, however, soon distanced himself from the cut-and-dried utilitarianism of the conservation movement; he began to think of nature as something more than "resources" requiring efficient management: nature provided a whole world of "outdoor culture" necessary to the revitalization of what had become an "iron civilization." Combining his Thoreauvian love of nature and his practical training, MacKaye formulated a plan

[4]John L. Thomas, "Lewis Mumford, Benton MacKaye, and the Regional Vision," in T. P. and A. C. Hughes, eds., *Lewis Mumford: Public Intellectual* (New York: Oxford University Press, 1990), pp. 80–81.

[5]Samue P. Hays, *Conservation and the Gospel of Efficiency: The Progressive Conservation Movement, 1890–1920* (Cambridge: Harvard University Press, 1959); also see Robert Wiebe, *The Search for Order, 1877–1920* (New York: Hill and Wang, 1967).

for the U.S. Department of Labor providing for the resettlement of depopulated rural areas.[6] Two years later, having recently left government service, MacKaye reconceived his Department of Labor plan in his article calling for the creation of an Appalachian Trail.[7]

With the publication of this essay, MacKaye gave concrete expression to what became the theme of Mumford's *Technics and Civilization*: the possibility of finding a synthesis of modern technics, regional economic development, and ecological thinking. MacKaye called the new technique "geo-technics," regional planning informed by an georegional sensibility. "Geo-technics" begins with awareness of land(scape): in his call for the development of an Appalachian Trail, MacKaye revealed his own passage *back* into an interior, continental realm, a wild(er) place suitable to the "mystery" and resonances of origin. The trail, he contended, permits one to pass from civilization *into* the wild, where, like Thoreau, one finds "it all . . . huge and real . . . the thing itself." Hiking suggests recreation, but not entertainment or diversion: recreation is "outdoor culture," an experience of re-creating self. As Paul Bryant explains, this is re-creation as the act of "reviving or creating over again some of the profound aspects of [hu]man nature."[8]

[6]Benton MacKaye, *Employment and Natural Resources* (Washington, D.C.: Government Printing Office, 1919), advocated the establishment of agricultural, forest, and mining communities on federal lands. By providing employment for returning veterans, these cooperative settlements would also reverse the migration into overcrowded urban areas while encouraging a kind of stewardship of the land under the federal government's auspices. The idea was to maintain federal land tenure while leasing the land to cooperatives, organized for the purpose of mining, lumbering, or agriculture. The settlements would be permitted and encouraged to find appropriate forms of "self-organization" along cooperative lines. The federal government would require harvesting practices in line with the national goal of conservation of resources; the government would also benefit from all profits extracted from the land yet would compensate settlers for depressed economic conditions by guaranteeing a minimal annual income for each family.

[7]Benton MacKaye, "An Appalachian Trail," *AIA Journal* 9 (October 1921): 325–330.

[8]Paul Bryant, "The Quality of the Day: The Achievement of Benton MacKaye" (Ph.D. dissertation, University of Illinois, 1965), p. 152; Benton

The Appalachian Trail was to serve as a catalyst for the experience of the wild and to become an expression of a rising awareness of the importance of wilderness. The trail proposal makes clear MacKaye's emphasis on encountering and entering into the land, which means first and foremost knowing the land. It also suggests the spiritual importance of accessible landscape, rural and wild. The trail was to be accessible as a common ground available to all citizens. As Mumford remarked in his introduction to MacKaye's book on regional planning, *The New Exploration*: "it was not merely . . . proposing a trail of such length that [distinguished MacKaye's plan; it was that] . . . he conceived of this trail as the backbone of a whole system of wild reservations and parks, link[ed] together by feeder trails into a grand system, to constitute a reservoir for maintaining the primeval and rural environment at their highest levels."[9] It is the systemic quality of MacKaye's idea that should attract our attention; more than a park that sets aside wildlands, MacKaye proposed a system to maintain wild and rural lands *in relation to* urbanized areas (see Figure 3). The trails link these different landscapes, physically and metaphorically bringing their different outlooks on life into common relation. Walking a trail is an active relationship to landscape because it is operative. This geographic passage MacKaye encouraged was metaphorical, leading us out of the city into the wild, and from the wild back to the city.

For the purposes of regional planning, the key linkage is between landscape and geography as a system. MacKaye understood the natural region as a geographic system: a natural pattern established by terrain and confirmed by waterflow that not only describes the natu-

---

MacKaye, "Outdoor Culture—The Philosophy of Through Trails," *Landscape Architecture* 17 (April 1927): 163–171; quoted in ibid. p. 126. Thoreau, *Cape Cod* (New York: Crowell, n.d.; reprint ed. New York: Apollo, 1961), p. 74.

[9]Mumford, "Introduction," in Benton MacKaye, *The New Exploration: A Philosophy of Regional Planning* (Urbana: University of Illinois Press, 1962; orig. pub. New York: Harcourt, Brace, 1928), p. xiv.

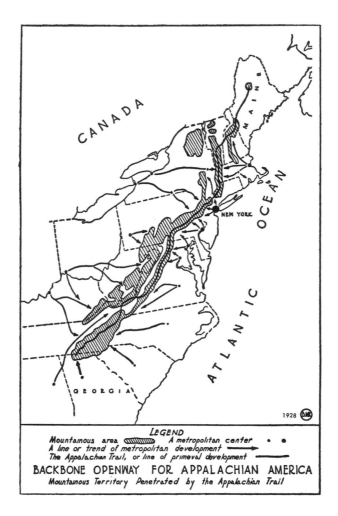

FIGURE 3. "Ecological Regionalism" I. MacKaye's trails are part of a regional and interregional system intended to put wild, rural, and urban geographies in relation to one another.

ral system but explains the movement of human populations and the growth of economies. Just as a river's drainage is shaped by the terrain, human activities follow the course of topography (see Figure 4). For MacKaye, planning began with exploration. As the old explorers followed the river systems into the heart of the continent, the new explorers must follow the flow of commodities into the heart

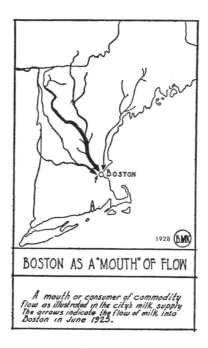

BOSTON AS A "MOUTH" OF FLOW

A mouth or consumer of commodity flow as illustrated in the city's milk supply. The arrows indicate the flow of milk into Boston in June 1925.

FIGURE 4. "Ecological Regionalism" II. MacKaye used nature—the flow of water as shaped by the land—as a metaphor for economy and demography. Milk supplies follow the river into the city.

of the "industrial wilderness": "the river or water stream is the guide to the *terra incognita* of the continent; the traffic or commodity stream is the guide to the *terra incognita* of industrial civilization."[10] Planning is founded on geographic exploration: tracing the stream of traffic, both population and commodities, from the urban hinterland—the "watershed" regions from which the raw materials and population are drawn—along "traffic streams" to their "conflux" in the metropolis (*NE*, p. 12). Planning should seek not only to regulate and control these flows of goods and people but to reconceive and reconfigure them.

MacKaye argued that the moral imperative for doing so arises from the effect of present patterns of development and growth: the commodity and population flows of "industrial civilization" are destroy-

---

[10]MacKaye, *New Exploration*, p. 28. Subsequently referred to as *NE*.

ing *both* the hinterland and the city. The hinterland is being systematically stripped of resources and populations; there forestry and agriculture have modeled themselves after mining, leaving devastation in their wake. Among the resources being destroyed (now as much as then) are what MacKaye called the "primeval" lands and the "indigenous cultures."[11] The primeval is what we now call the "wilderness": lands MacKaye recognized as "the environment of life's sources, of the common living-ground of all mankind"(*NE*, p. 56). The "indigenous cultures" correspond to what Mumford called "regional cultures": the geographically distinctive and diverse cultures of the provinces. The economic imperative that is destroying the regions also affects the large urban centers. A "true city" has "definite boundaries and an individual integrity" (*NE*, p. 171). This integrity is diminished as economic growth entails greater and greater urban sprawl. The city has become a "glacier": "it is spreading, unthinking, ruthless. Its substance consists of tenements, bungalows, stores, factories, billboards, filling-stations, eating-stands, and other structures whose individual hideousness [is only exceeded by their] collective haphazardness" (*NE*, p. 160).

MacKaye invited readers us to understand this process of urban growth in geographic terms: his metaphor is a water system. As the terminus of inflowing streams of ever greater volume of goods and population, the urban "reservoir" has been steadily increasing in size. When it reaches overflowing, the consequence is a "backflow." Suburbanization and the diffusion of industry may be understood as the consequent "flow of population" and "flow of industrial plants" out of the city (see Figure 5).

Any genuine program of regional planning must come to terms with these geographic movements. MacKaye suggested that the first step should be to contain the metropolis. He proposed linking a series of natural reservoirs—mountain crestlines, escarpments, canyons, and wetlands—as "levees" to inhibit the "metropolitan flow"

[11]MacKaye's discussion of the "primeval," (see for example, *NE*, pp. 138–140) makes clear his appreciation of wilderness. Unfortunately for the sense of our present meaning, MacKaye used *wilderness* in the purely negative sense of barbaric and chaotic; thus he referred (p. 24) to the "wilderness of civilization—[a] jungle of industrialism."

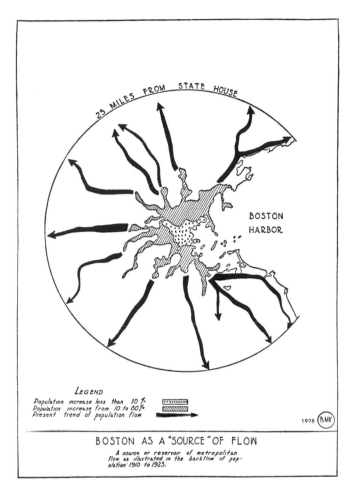

FIGURE 5. "Ecological Regionalism" III. The metropolitan center is the "source" of population "outflow."

(*NE*, p. 195) (see Figure 6). Yet he realized that even the completion of a natural system of dikes (such as Frederick Law Olmsted's "emerald necklace" of regional parks that encircle greater Boston) would not in itself contain metropolitan sprawl. One problem MacKaye identified is that the levees will have to be breached by traffic streams—specifically, railroads and highways. These are the lines of vulnerability. To create effective barriers to urban sprawl, these streams of traffic through the levee system must be carefully controlled. MacKaye was most concerned with highways because he recognized that they

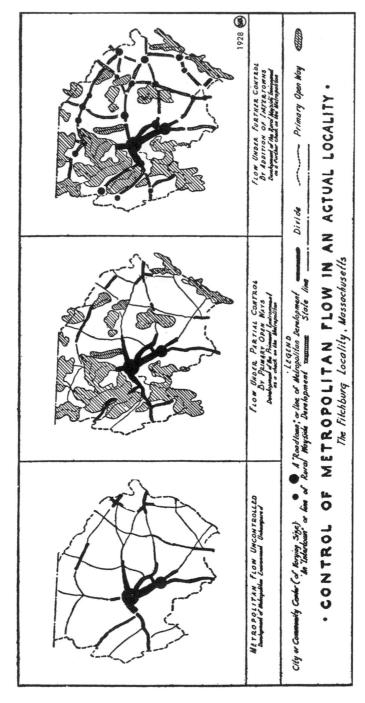

FIGURE 6. "Ecological Regionalism" IV. The solution to population "outflow". Protected green lands form an "intertown" network of rural and wild places.

would be increasingly relied on for transportation and because automobile and truck travel is more flexible than the railroad. Because it may be stopped and started frequently, the motorized vehicle running on streets and highways tends toward dispersion of facilities. Consequently, along highways a new kind of built environment is created, a variant of urban sprawl that MacKaye found visually abhorrent: the commercial strip development, or "roadtown." And where commercial strips grow up, factories and residential areas will follow. Consequently, if the damming of metropolis is to stand a chance of succeeding, land use adjacent to highway right-of-ways must be controlled. In this context MacKaye offered one of his most original planning proposals: the construction of "townless highways" that bypass urban areas;[12] their purpose is not only to accommodate traffic flow but to minimize the effects of the heavy traffic on town and city centers and, critically, to virtually eliminate urban sprawl alongside the new highways.[13] This addresses the problem of urban "backflow."

MacKaye understood that containing the metropolitan "backflow" through a system of natural preserves breached only by strictly controlled limited access highways is only a temporary solution. It does not address the fundamental problem of the growing strength of the geo-economic forces that have necessitated the continual expansion of metropolis. Regardless of the strength of the system of levees, the continuing inflow of population and commodities will create tremendous pressure for expansion. Even if constructed, the dams will not hold indefinitely. Or to put this into other terms: unless the economic imperatives are altered, the concentration of economic power, population, and capital in the metropolitan centers will overwhelm any attempt to contain the growth and spread of those centers.

[12]The basic proposal was to construct what traffic engineers call "limited access" highways (the access is controlled by constructing under- and overpasses and limiting access through interchanges) similar in design to the parkways of the 1920s. Like the parkways and unlike the present interstate system, development around the highway would be strictly regulated—and largely prohibited.

[13]For a detailed account of his highway proposal, see "The Townless Highway," *New Republic*, 12 March 1930, pp. 93–95.

Accordingly, MacKaye proposed to reconfigure the patterns of urban growth and development through a comprehensive regional economic planning directed at redistributing population and redesigning
the crucial commodity "flows." Industry would be located nearer the
sources of raw materials, in accord with the natural pattern of resource flow in the "industrial watershed" (*NE*, p. 42). Planners, if
they are to be a true "geo-technicians," must carry out a site analysis
in the context of this natural pattern.[14] And if industry is relocated,
population will follow: regional cities, "groupings of small cities or
villages within an area," will draw economically and culturally upon
these industrial and natural watersheds and manage to combine "the
attributes of the natural region and the true city" (*NE*, pp. 165–166).

MacKaye's (and the RPAA's) paradigm for this kind of regional economic development was a significant part of his Appalachian Trail
proposal. The most salient characteristic of the proposal was the linkage MacKaye established between rural (economic) redevelopment
and cultural regionalism. More than a "refuge from the scramble of
every-day worldly life," the trail would thus provide the opportunity
for settlement, the means of establishing a "new rural population."
The economics of MacKaye's plan was based on planned camps or
settlements near the trail. In providing employment for inhabitants
of the region, an effort to "counter migration from city to country,"[15],
the camps would be the center of economic activities, particularly
sustainable forestry. But they would also be hostels for hikers, providing another economic base. MacKaye argued that rural areas must
have viable economies, and at the same time the most important,
long-term economic asset is a sustainable natural ecology. In other
words, a rural region stands to benefit economically if its wilderness
attributes and ecological balances are sustained. In the case of Appalachia, the entire habitat, trail and camp settlements, would provide
an economic alternative to the unsustainable forest industry prevalent then (and today) and the "recreation" provided by roadside camps

[14]Benton MacKaye, "The New Exploration: Charting the Industrial Wilderness," Survey Graphic 7 (May 1925): 153-157, 192, 194; reprinted in
Carl Sussman, ed., *Planning the Fourth Migration*, (Cambridge: MIT Press,
1976), pp. 94–110.

[15]MacKaye, "An Appalachian Trail," pp. 327, 329.

and "motorslums" which had already begun to dot the countryside. At the same time the trail would provide a focal point for sustaining cultural regionalism within the context of the preservation and appreciation of wildlands. In a sense, MacKaye was arguing in favor of the conservationist position in their debate with preservationists about the disposal of America's wildlands. His central point was that "multiple use" is possible—but only if it is based on sustainable harvests of renewable resources and *designed*, that is, regionally *planned* to be geographically compatible with preservation of wildlands.

In the broadest sense, then, regional planning could reshape the relationship between culture and nature. MacKaye's contribution was a concept of planned regional development that made it possible to use geographical resources without destroying them: a diversified kind of economy made compatible with the preservation of a region's natural character. This extended Mumford's hope for uniting economic function and aesthetic appreciation. In MacKaye's work, as in Mumford's, regionalism is not purely aesthetic, for the aesthetic extends to the shaping of a geo-economy. A productive economy that grows from geographic possibilities is as much a function of place as is a system of trails that lead us imaginatively back to the wild.

This sense of geographic regionalism moves toward the restoration of what R. F. Dasmann calls an "ecosystem culture." For it would develop the natural and cultural resources of the region; as such, Dasmann contrasts it to a "biosphere culture" based on urban "supra-ecosystems" that focus human and natural energy on the interchange of remote and unrelated ecosystems.[16] Regionalism would tie local and regional cultures and economies to an ecology growing out of the particularities of the natural region while engendering community and regional consciousness. This does not mean that MacKaye or

[16]R. F. Dasmann, quoted in Gary Snyder, *The Old Ways* (San Francisco: City Lights, 1977); see also R. F. Dasmann, *Environmental Conservation* (New York: John Wiley, 1984), esp. pp. 436–442. For clarification, Dasmann defines *biosphere* as the part of the planet where life exists, the sum of all ecosystems; the ecosystem is a subdivision of the biosphere, a "community" of plants, animals, microorganisms, and the air, water, and soil that supports them. An ecosystem can often be defined as the outcome of a "process of succession" that leads to a stable "climax community."

Mumford advocated the creation of economically self-sufficient regions, and it certainly does not suggest the return of an agrarian economy, the possibility of which Mumford called "an absurdity."[17] But regionalism did suggest the restructuring of trade and population, such that a

> state of economic balance [would be created] . . . in which the population of a region will be redistributed with respect to its fundamental resources, in which agriculture, the extractive industries, manufacture and trade will be coordinated, in which the size of cities will be proportioned to open spaces and recreation areas and *placed in sound working relation with the countryside itself.*"[18]

❖

MacKaye confirmed Mumford's conception of a comprehensive regional planning and strengthened his insistence on the fundamental importance of regionalism as the framework for the RPAA's mission. MacKaye also contributed two critical ideas about planning. First, planning is a method of designing the built environment that aims not at replicating nature but at incorporating organic principles of design. MacKaye understood this goal as part of a process of entering into an imaginative relation to nature. Second, he saw planning as a way to ground economy, an otherwise self-contained and highly abstract discipline, in geography. Planning can help define an ecologically appropriate economy by translating "a sense of place" into a set of geo-economic practices appropriate to particular regional settings.

These design principles help define a holistic planning that brings culture and nature into relation. The principles also share a common root in MacKaye's conception of planning as an activity—a way of

---

[17]Mumford was well aware of the danger of chauvinism inherent in many attempts at economic and cultural self-sufficiency: "the dream of autarchy is merely a military dodge for putting a population in a state of mind appropriate to war." Furthermore, he added, "no region is rich enough or varied enough to supply all the ingredients of our present civilization" (*C of C*, p. 345).

[18]Mumford, "The Theory and Practice of Regionalism," *Sociological Review* 20 (January 1928): 18–33, esp. 25–26, emphasis added.

being in the world, of entering into environment. I think this is what he meant by "exploration," an experiential basis for planning. There-fore, while he admitted the necessity of developing the technical expertise necessary to all kinds of planning—from designing road-ways to parks and towns—MacKaye insisted that planning begins with an exploration of the environment. In *The Culture of Cities*, Mumford showed that this geographic exploration is also a means of self-development. Infants understand themselves, in large part, in relation to their immediate environment. They define themselves both against and with the environment. This process, usually truncated by social conditioning during the school age, should be continued, Mumford's argued, as a part of the formal and informal educational process: "the systematic contact with the environment should broaden out until it includes the furthest horizon of mountain top and sea" (*C of C*, pp. 383–384).[19] In this way all citizens would share in a training based on a conception of subjectivity that encounters ("ex-plores") rather than negates the world ("environment"). And all plan-ners would share in this process as well.

Thus planning based on exploration is potentially a democratic idea. For despite its necessary concern with technical expertise and scientific discipline (aspects that lock out the citizenry), planning can be rooted in a commonly shared human activity: the capacity to explore. While neither Mumford nor MacKaye emphasized citi-zen participation in planning, their intention is clear: techniques such as regional planning must reflect a sense of place and a sense of collective responsibility, and these qualities can be attained, in some measure, through direct experience of the natural world. And this experience is a necessary common ground for the exercise of a socially and ecologically responsible planning practice. Mumford and MacKaye challenged the predominant conception of planning as a narrow specialty. In effect, they understood that we are faced with the task of reexamining and reordering the relationship be-

[19]Robert Dickinson takes up this challenge in *Regional Ecology: The Study of Man's Environment* (New York: John Wiley, 1970) with a concrete pro-posal for a school curriculum (pp. 165ff).

tween technique and experience and between expertise and participation.

❖

In *The New Exploration* MacKaye stated that the key to regional
planning is the perfection of an "applied science" that enables "charting of possible [or potential] facts lying within [nature's pattern]"
(*NE*, p. 30). Topography is the key to understanding natural patterns. Consider, for example, the engineer who designs a switchback
as a way to cross the Rocky Mountains:

> he does not really *plan* the switchback, he *finds* it—out in the
> mountains amid the facts and laws of nature. He does not create
> his own plan, he discovers nature's plan; he reveals a hidden po-
> tentiality which nature's laws allow. Likewise with the hydro engi-
> neer who harnesses Niagara Falls and develops from a flow of water
> a flow of electric power. . . . So also with the agriculturalist who
> develops from a region's farming crops a steady food supply. And
> with the city planner who controls the flow of a local population.
> Each one of these is a type of engineer, a man who *finds* rather
> than *plans* a region's best development: one who builds on the
> actualities disclosed by exploration. (*NE*, pp. 33–34)

By combining the efforts of these specialists, a regional plan may be
developed that permits the "highest use" to be "secured from [a
region's] natural resources as a whole." The possibility of using natural resources in an environmentally compatible manner is inherent
in nature's design. In its most profound sense, environmental planning is not about limiting human destruction of the environment or
providing "natural preserves" that may stand apart from "human use"
areas: it is about "exploring" the terrain and permitting what we find
to suggest how we use the land. The engineer, or "geo-technician,"
who plans the mountain switchback uses the natural geography to
plan; he does not ignore the geography and impose a use on nature.
The same thinking, MacKaye reasoned, must be applied to all human uses of the land. We must consider the resources and design of the
region when planning its cities, industries, farms. To the sensitive and
patient observer, the geography suggests the best use, and it will be a
purpose that is compatible with the natural integrity of the region.

MacKaye conceived of the region, then, as a natural *design* clearly expressed in its topographical features. This idea was powerful because it gave MacKaye a means of placing culture (economy) in relation to nature (region). His most important insistence was on the possibility of finding principles of design that reflect nature. Furthermore, MacKaye understood that the basis of building is more than aesthetic, that it concerns more than either architecture or even the visual relation to the land. The built environment is in relation to "commodity flows," which is to say it concerns the (economic) structures of production, transportation, and commerce. More than laying out visual criteria, the principle of organic design contributes an important perspective for reinventing economy: it calls into question the imperative of agglomeration. If the concentrations of industry and population in megacities could be reconfigured, MacKaye argued, we stand the chance of creating a new kind of urban environment, one in harmony with regional design. To escape the metropolitan agglomeration means building anew, and in this respect MacKaye's regionalism is linked to the garden city movement.

In practical terms the emphasis on nature as design had an important effect on the RPAA's conception of technology and technological modernization. It helps explain why the RPAA proclaimed the environmental compatibility of automobiles, limited access highways, large power dams, and electric power lines: they would be planned in accordance with natural regional patterns. Unfortunately, as we know, these new technologies would be used to spread urbanization outward from the great metropoles into rural regions. And the detrimental environmental effects we now attribute to such technologies was to be greatly exacerbated by those whose planning practice, unlike that of MacKaye and the RPAA, was not informed by concern with natural design and regional balance.

The work of Mumford, MacKaye, and the RPAA has often been attacked as an instance of a perceived anti-urban bias in American culture. Some critics have argued that the implications of RPAA planning were romantic in the pejorative sense that the group failed to address the realities of urban American life.[20] These critics interpreted regionalism as a retreat into an idealized pastoralism reminiscent of the 19th-century popularization of romantic nature.

The Hudson River School of landscape painting originated the American version of pastoralism, but its most concrete expression was found in the architecture of Andrew Jackson Davis and the landscape architecture of Andrew Jackson Downing. These 19th-century romantics fused culture and nature into a single representation, usually an idealized, pastoral landscape that mediated the polarities of art and nature, cultivation and wilderness. Landscape painters such as Frederic Church and Thomas Cole depicted a rural landscape—a pleasant example of human transformation of the land—enveloped within a more comprehensive setting of picturesque and sublime nature. These paintings encompass a humanized rural landscape within the context of a wild and irregular "Nature." Downing's work in landscape architecture represents the best articulation of this ideal relation between wild and cultivated nature. Borrowing a set of techniques from English gardening, he designed landscapes that enveloped the viewer alternately in the "beautiful" (a tame, pastoral nature achieved by emphasizing the roundness and regularity of space enhanced by soft deciduous trees planted singly or in uniform groupings) and the "picturesque" (a wild, irregular nature created by accentuating broken and abrupt surfaces such as rocky outcroppings surrounded by dense growths of irregular evergreen trees). Downing focused his work in landscape and architecture on the rural world, producing villas for the "natural" Jeffersonian aristocrats and cottages for sturdy yeomen. In effect he united the social ideals of Jeffersonian republicanism with an aesthetics drawn from English romanticism.[21] Perhaps it was inevitable that this synthesis of repub-

[20]See footnote 23.

[21]This important attempt to "Americanize" the natural landscape garden is precisely the reason Downing deserves special attention in the history of American landscape. Though we now know that the "English garden," or naturalistic landscape, in America predates Downing (see Paula Dietz, "A Historic Garden is Recovered From the Rough," *New York Times*, 25 June 1992), having turned up on a South Carolina plantation as early as 1730, Downing was the first to publicize the naturalistic landscape as paradigm and to make it compatible with American social ideals.

lican rural virtues and romantic rural sensibility would be viewed as a key construction of American culture; Lucia and Morton White, whose influential book *The Intellectual Versus the City* appeared in 1962, made it their central thesis. Jeffersonian republicanism and romantic pastoralism, first brought together by Downing, appeared to the Whites as a cultural motif related to the postindustrial flight from the city and the ideal of suburban life; it suggested a deeply rooted American cultural bias against the city. In *The Machine in the Garden*, Leo Marx made an important contribution to this thesis by arguing that Americans were preoccupied by a simplistic pastoralism, an attempt to find or create a humanized natural landscape that encompassed technological innovation and yet obscured the impact of the machine. The result: America was ill equipped to deal with the realities of an industrialized, mechanized, urban society.

One of the most important links in this agrarian thesis is the work of Frederick Law Olmsted. He was perceived as the key transitional figure: the most significant designer to link Jeffersonian agrarianism to the modern industrial city. Yet when recent historians, beginning with Albert Fein, examine Olmsted's work, they find that far from an "agrarian" bent on dissolving the city—a 19th-century Frank Lloyd Wright planning a diffused "city," with each family secure on its own small homestead—Olmsted was an urban planner passionately concerned with the revitalization of cities. Certainly, Olmsted applied Downing's landscape principles to the planning of parks, notably Manhattan's Central Park; and he was concerned with planning suburbs. But unlike Downing, Olmsted believed that the central task of the landscape architect was to reverse the decline of the polis as manifested in the commercial cities of his day. Only when natural parks containing picturesque landscapes and tree-lined "parkways" infused the city with the grace and variation of nature would the commercial city become a civilized place in which to live. Thus, while Olmsted shared Downing's aesthetic, Olmsted applied it in a decidedly urban context.

Olmsted's work led in two directions. One was the City Beautiful movement, which provided many cities with parkland and tree-lined boulevards. Daniel Burnham, the leader of the City Beautiful movement, built on Olmsted's urban vision, combining its pastoral elements, civic concern, and anticommercial spirit with an emphasis on the monumental. The City Beautiful movement's concern with designing the city around monumental buildings and public spaces may well have been, as Mumford argued in *The Culture of Cities*, a worship of imperial power, but it offered a civic vision and an alternative to the purely commercial city. The other direction in which Olmsted's work evolved was the building of residential suburbs. Olmsted's plan for Riverside outside of Chicago became a model for the construction of picturesque residential subdivisions. These suburbs were set apart from the existing urban centers, but they were intended to be a part of the urban experience. They were not to function as exurban (or rural) estates, though, again, they incorporated many of the principles of landscape design that Downing had applied to rural settings. For Olmsted, unlike Downing, wanted to create pastoral residential suburbs that would "civilize" and elevate urban life. Olmsted understood that the commercial city was changing under the pressure of industrialization; he was anxious that these changes reflect aesthetic principles. He understood that the improvements in transportation and communication were creating a functionally differentiated metropolis—a city of related working parts. Olmsted's parks and suburbs were an attempt to incorporate a pastoral aesthetic *as a part of metropolis.*

Olmsted's achievement was important but limited. His great parks showed that the pastoral aesthetic could find an appropriate application to modern urban areas. Most important for my purpose is the effect Olmsted's work had on agrarian pastoralism: it shattered Downing's association of the pastoral aesthetic with republican agrarianism. But Olmsted certainly did not make pastoralism a progressive social force. After all, for all his good work in creating magnificent urban parks that demonstrated that cities need not do away with natural landscapes, his pastoral suburbs reinforced the class segregation and sexual differentiation (work in the city for men, staying at home in the suburbs for women) of 19th-century middle-class culture. Though Olmsted's suburbs were a part of the evolving metropolis, they were designed to create a separate domestic sphere as a function of a culture that believed in rigid class distinctions and in a sexual division of labor.

Still, that Olmsted's work was oriented toward urban development is a strong indication that the idea of a governing *agrarian* myth is far too simplistic. This is not to say that forms of agrarianism did not survive into the 20th century (or, for that matter, that some were without any social value). In the 1920s and 1930s, back-to-the-landers drew their inspiration from Ralph Borsodi (*This Ugly Civilization*, 1929, and *Flight from the City*, 1933). His Suffern, New York, experiment, a "town" of self-sufficient homesteads, was a (self-admitted) retreat from urban society. Yet by arguing that small-scale, modern machinery could supplement mass production, Borsodi make a valuable contribution to community planning. Borsodi's agrarianism may have been unrealistic as a paradigm for social reorganization, but what he had to say about the scale and organization of production is important. Furthermore, by changing the context of his critique from the individual homestead to the whole *community*, we can adapt Borsodi's ideas about overcoming the immense scale and impersonal organization of mass production. This is exactly what Paul and Percival Goodman did in their visionary proposal *Communitas*, in which they adjusted small-scale production to a community-centered urban life.

At the same time, some agrarians of this period were reactionaries who used agrarianism to promote a static and hierarchial social order. The southern agrarians belong to this category. The group's manifesto, *I'll Take My Stand*, was "anti-progressive," "anti-rationalist," and "anti-humanist." The Old South, according to Allen Tate in his book on the Confederate President Jefferson Davis, represented "the forlorn hope, of conservative Fundamentalist Christianity and of civilization based on agrarian, class rule, in the European sense." "The issue," he added, "was class rule and religion *versus* democracy and science."[22] The southern agrarians preferred a class-stratified, Eurocentric, "classical," Anglo-Catholic civilization based on an agrarian economy. They were advocates of reviving an ancient régime that had never existed in the United States. The agrarians negated "in-

---

[22]Twelve Southerners, *I'll Take My Stand: The South and the Agrarian Tradition* (reprint ed. New York: Harper and Row, 1962); Allen Tate, *Jefferson Davis: His Rise and Fall* (New York: Milton, Balch, 1929), p. 87.

dustrial civilization" so completely that they were left in the position of being hopeless reactionaries. The fundamental difference between Mumford and Tate was the content of their cultural critique and philosophical assumptions. The agrarians were idealists; they based their critique of culture on a transcendent and static ideality, as contrasted with Mumford's dynamic, process-oriented pragmatism.[23]

In any case the emergence of Olmsted's work marked the decline of agrarian pastoralism, which was replaced by what James Machor calls "urban pastoralism." Olmsted was the seminal figure in this widespread cultural movement, which included many late-19th-century writers and critics. "Urban pastoralism" developed an aesthetic based on a synthesis between "nature" and the urban environment. This new version of the "middle landscape" (the old version was a humanized nature in a pastoral or agricultural setting) was applied to both urban parks and residential suburbs; it became for the culture what John Stilgoe has referred to as a "borderland" between and amidst the city and the natural world. It is against this cultural norm of a pastoral-urban synthesis that the RPAA's conception of regional planning and Mumford's regionalism must be considered.[24]

[23]The RPAA held a conference hosted by Stringfellow Barr at the University of Virginia in 1932. But most of the agrarian group repudiated the RPAA for supporting industrialization of the South. See John L. Stewart, *The Burden of Time* (Princeton: Princeton University Press, 1965); see also R. A. Lawson, *The Failure of Independent Liberalism* (New York: Putnam, 1971), pp. 135–141, on the related thinker Herbert Agar and the "Distributist" movement of the 1930s.

[24]Lucia and Morton White, *The Intellectual Versus the City* (New York: Norton, 1962); and Leo Marx, *The Machine in the Garden* (New York: Oxford University Press, 1964), are the basic interpretative texts. While disputing the connection between Jeffersonianism and romanticism, John Stilgoe's comprehensive study of the suburban cultural ideal, *Borderland* (New Haven: Yale University Press, 1988) builds on the anti-urban thesis by locating the cultural impulse of suburbanization in the quest for "country." See also Peter J. Schmitt, *Back to Nature* (New York: Oxford University Press, 1969) for his exposé of the sentimentalized and commercialized "nature" of suburbanites; and Kenneth T. Jackson on "romantic suburbs" in *Crabgrass Frontier* (New York: Oxford University Press, 1985), pp. 73–86. Wright's hope for replacing the city with the auto-dominated suburb is apparent in *The Disappearing City* (New York: W. F. Payson, 1932); both his

Certainly, the RPAA appropriated aspects of the tradition of park and suburb building.[25] In an article for *Harper's* magazine, Mumford sympathized with the suburban impulse, arguing that the "suburb is an attempt to recapture the [natural] environment which the big city

---

design and his conception of the suburb as a diffused city stand as a good contrast to Olmsted. In "The City in Agrarian Ideology and Frank Lloyd Wright," in *The American City: From the Civil War to the New Deal*, trans. B. L. LaPenta (Cambridge: MIT Press, 1979), Giorgio Ciucci correctly sees Wright's planning proposals as a resurgence of "the [arcadian] dream of regaining a rural existence," though he overstates the breadth and simplifies the meaning of an "arcadian myth." He succumbs, I think, to the tendency to collapse the distinctions among "pastoralists."

For his critique of simple pastoralism and documenting the "urban pastoralism," I am particularly indebted to James L. Machor, *Pastoral Cities* (Madison: University of Wisconsin Press, 1987). I also want to credit David Schuyler, *The New Urban Landscape* (Baltimore: Johns Hopkins University Press, 1986) for pointing out the ambivalence of American attitudes toward the city, the distinction between republican agrarianism and romantic pastoralism, and the continuing commitment to cities of those such as Olmsted, who intended to incorporate romantic scenic values into urban life.

On Downing, see his *Treatise on the Theory and Practice of Landscape Gardening* (New York: Wiley and Putnam, 1844); George B. Tatum, "A. J. Downing: Arbiter of American Taste," (Ph.D. dissertation, Princeton, 1949); W. G. Jackson, "First Interpreter of American Beauty," *Landscape* 1 (Winter 1952): 11–18.

On Olmsted, see a collection of his documents edited by A. Fein: *Landscape into Cityscape* (Ithaca: Cornell University Press, 1968); and his essay "The Structure of Cities: A Historical View," in S. B. Sutton, ed., *Civilizing American Cities* (Cambridge: MIT Press, 1971), in which Olmsted maintains that urban growth is inevitable and suggests planning a functionally differentiated metropolis with parks serving as accessible countryside. See also A. Fein, *F. L. Olmsted and the Environmental Tradition* (New York: George Braziller, 1972); and Laura Wood Roper, *FLO* (Baltimore: Johns Hopkins University Press, 1973).

On the regressive social ideals that attracted the middle class to suburbs, see Robert Fishman, *Bourgeois Utopias* (New York: Basic Books, 1987).

[25]Italian architectural historian Francesco Dal Co, "From Parks to the Region," in Giorgio Ciucci et al., *The American City: From the Civil War to the New Deal*, trans. B. L. LaPenta (Cambridge: MIT Press, 1979), has argued that the RPAA belongs squarely within the pastoral tradition. He sug-

. . . has wiped out within its own borders."[26] Olmsted's parks and romantic suburbs were "real social mutations in urban form" that helped to "break up the clotted urban massing of the great metropolis" (*C of C*, p. 221). But Mumford's praise indicated that it was not pastoralism but organicism and regionalism that Mumford appreciated in Olmsted's planning practice. In *The Culture of Cities*, Mumford appraised Olmsted's use of naturalism and his attention to *place*:

> [Olmsted] took the capricious naturalism of the earlier lovers of dead trees and broken branches, and the meaningless irregularities of the Jardin Anglais, and created new harmonies, based upon

---

gests his own formulation of urban pastoralism, positing a line of cultural development that begins with the Transcendentalists and proceeds through Olmsted and the parks movement to Progressivism and the RPAA. The mission of the RPAA, as Dal Co sees it, was to impose a pastoral sensibility—derived from the Transcendentalist desire to infuse civilization with a static nature and manifested in a set of "ahistorical ideals"—on a conception of a planned urban complex that could suit the capitalist agenda of infrastructural modernization. So the RPAA stood between, and worked out a viable linkage of, the pastoral ideal and the "neo-technic" order that had been a feature of American reform since Edward Bellamy presented his vision of technocratic utopia in *Looking Backward* (Boston: Houghton, Mifflin, 1889) and the use of planning and electrical technology had been demonstrated in Chicago at the World's Columbian Exposition of 1893. The RPAA, then, constructed an urban pastoral myth that served to deflect criticism from the "real" (i.e., historically constructed) development of American urban areas. The impetus, Dal Co concludes, seems to have been the "attempt of the bourgeois spirit to substitute ethics for politics" by creating a "world of values" derived from an "aristocratic intellectual heritage"—the 19th-century revival of "civic" concern. Dal Co's criticism has much to recommend it, particularly his insistence on the failure of the RPAA to develop its ideas in a political context. Mumford's lack of a carefully articulated politics left him open to this charge. But it is misleading to approach the RPAA as a cover for capitalism; American capitalism, as we shall see, had little use for the RPAA, preferring laissez-faire to systematic regional planning. Furthermore, and this is the essential point, Dal Co misunderstands Mumford's attempt to subordinate pastoralism to organicism and apply it to a larger project of regional reconstruction. Also see Edward Relph, *The Modern Urban Landscape* (Baltimore: Johns Hopkins University Press, 1987).

[26]Mumford, "The Intolerable City: Must It Keep on Growing?" *Harper's* 152 (February 1926): 283–293, quote on 289.

a closer study of the lay of the land and of native plant formations. He was not afraid to use the commonest wild flowers and native shrubs in his planting; and he followed the contour lines in the laying out of paths and roads. . . . His plans were animated by technical intelligence and civic foresight. (*C of C*, p. 221)

In Mumford's reading Olmsted was interested in something beyond nature as an aesthetic amenity. Experiencing "nature" in one of Olmsted's parks encourages a relation to place with both aesthetic and social aspects. In aesthetic terms the attention to "native plant formations," to "common" flowers, and to the "contour lines" of the land helps define the relation between the park and the natural region it should evoke. In social terms the park provides a commons that encourages the development of "civic" values.

Similarly, the romantic suburb puts residence in relation to a naturalistic setting. But the suburb could not transform the way of life of the modern urban dweller. Rather, it reinforced the structural differentiation of space inherited from the industrial city. To respond to this problem, the RPAA clearly distinguished garden cities from suburbs. The RPAA's very existence was predicated on the possibility of urban and social *reconstruction*. Garden cities were conceived not as bedroom retreats from urban life but as a means of recontextualizing urban life. They were planned as whole environments, encompassing within a relatively contained area work, residence, and commerce. And they were meant for all social classes. In effect, the RPAA hoped to appropriate aspects of the pastoral aesthetic, but also to recast it to encompass the group's larger social and cultural goals.

CHAPTER 7

"Dinosaur Cities"

In his Appalachian Trail proposal, MacKaye synthesized regionalism and regional planning, establishing a useful model that helped Mumford conceive the RPAA.[1] MacKaye's work confirmed Mumford's understanding that the recovery of nature and a more balanced kind of urban development required regional planning. But at the same time, Mumford understood that regional planning alone was insufficient to bring about the kind of social and cultural change that he envisioned. The RPAA, he felt, should not be limited to discussions of "policies" defined by experts and implemented by the state. The RPAA must be part of a wider regionalist perspective.

It is not surprising, then, that shortly after the RPAA was organized, Mumford expressed some misgivings: privately he thought that other than MacKaye none of the RPAA members was able to conceptualize the "social and civic side" of the garden city. Doubtful of the value of planning in isolation from the *revisioning* of cul-

---

[1]MacKaye's article on the trail was published in 1921, two years before Mumford and Stein met to found the RPAA.

ture and society, Mumford believed that "we must start a regional movement in America before we can have regional planning."[2]

Despite his doubts, Mumford thought that the RPAA presented an opportunity to develop a planning practice informed by regionalist design. Furthermore, the RPAA seemed to Mumford to be a response to the historical moment: the opportunity to reconstruct what Geddes had called "paleotechnic" civilization, the technological and urban forms of the 19th century's industrial revolution. This opportunity was framed, in part, by the development of the new technical-professional disciplines. By the 1920s urban and regional planning had fully emerged from the panoply of late-19th-century charitable causes associated with social reform. The National Conference on City Planning (NCCP) had become an important forum for planning professionals, and in the 1920s several new regional planning commissions had been formed. The concern of this new discourse of regional and urban planning was to facilitate and shape the emergence of a new form of city. One of the most important goals of the RPAA was to influence this discussion, largely by extending its scope to related issues, including housing, industrial and economic planning, and especially conservation of the natural world.

In its working agenda the RPAA reflected the planning tools and social principles developed by Stein and his colleagues as well as the regionalist input of Mumford and MacKaye. These included: (1) channeling all new development into planned cities; (2) making new housing more widely available by significantly lowering land and construction costs; (3) creating private-public partnerships for assembling large tracts of land and planning new garden towns and cities; (4) careful planning of new urban areas to include both coherent urban centers and cohesive neighborhoods; (5) constructing limited access highways to meet transportation needs and strictly limiting development between towns on transportation routes; (6) situating more industries in rural locations nearer sources of raw materials; (7) developing indigenous (regional) sources of energy (e.g., hydro-

[2]Quoted in John L. Thomas, "Lewis Mumford, Benton MacKaye, and the Regional Vision," in T. P. and A. C. Hughes, eds., *Lewis Mumford: Public Intellectual* (Oxford: Oxford University Press, 1990), p. 81.

electric dams) as a basis for economic development; and (8) providing significant agricultural and recreational green space adjacent to all new *and existing* urban areas.

▦

Though it was a small and informal group held together largely by friendship, the RPAA had its divergences. As we have seen, Mumford and MacKaye were the two regionalist voices. The other members had emerged from the progressive housing movement; led by Clarence Stein and Frederick Ackerman, they were reformers who built on a tradition of social architecture.

The difference between the regionalists and progressive reformers can be seen in their divergent interpretations of Howard's original proposal for garden cities. Mumford endorsed Howard's plan for communal ownership of land and municipal services. He understood that communal ownership could become a basis for creating a local economy and community institutions directly responsible to the public. However, Stein's failure to endorse a significant but radical facet of the garden city idea revealed his—and in general the RPAA's—moderately reformist position.

Before the RPAA, Stein had been associated with a movement for "community planning" that combined techniques for large-scale residential construction with a vision of neighborhood cohesiveness. "Community planning" bridged the housing and planning movements; an early example of socially conscious planning, it emerged in response to the severe housing shortages experienced by the working class of the nation's growing cities. But the movement first drew widespread attention during the debate over federally subsidized housing during World War I. After the United States announced its intention to enter the war, Ackerman and, to a lesser extent, Stein became prominent supporters of government intervention in the housing market. They promoted "community planning" in the pages of the *AIA Journal* and finally helped convince the federal government to build "community" housing projects for workers in defense industries. This provided the opportunity to construct demonstration projects before the close of the

War brought the federal government's brief episode in the housing business to an end.

❖

Beginning in the late 19th century, North American "town planning" had been the province of aesthetically oriented architects and landscape architects who produced the "City Beautiful": broad avenues terminating in vistas dominated by impressive civic structures in the beaux arts style. Spurred by the construction of the dazzling, electrically illuminated White City of Chicago's 1893 World's Fair, the City Beautiful found its suburban parallel in the planning of *picturesque* subdivisions—large lots, winding drives, and pastoral beauty—for the upper middle class. The early "town planners" concerned themselves with providing suitable residential and commercial environments in the emerging downtowns and suburban zones for the well-to-do, environments that unfortunately functioned to conceal from the powerful both the wrenching transformations of urban life and the overly commercial character of the culture they had created. The gray area between downtown and the fashionable suburbs—the newly emerging, sprawling factory districts, slums, and residential quarters for the working class—received little attention unless there was a proposal for slum clearance to facilitate construction of a parkway designed to "integrate" downtown with the new suburbs. This gray zone, the area of the most dramatic industrial growth, became the chief concern of the new City Useful movement that began about 1910. City Useful planners directed their energies to the development of an infrastructure necessary for "efficient" movement of goods and people. This agenda led these self-proclaimed "scientific planners" to redefine the city as a modern metropolis composed of functionally variegated but structurally integrated sectors: sprawling manufacturing districts, central downtown business districts, and suburban residential areas. The planners directed little thought toward the problem of working-class housing. Indeed, the modern metropolis posed a direct threat to the integrated, mixed-use neighborhoods of the old center city, which were increasingly relegated to commercial purposes. Housing would be located elsewhere.[3] For the middle class (the upper 20% of the income brackets), this meant the development of amenable urban and suburban residential districts. For the working class, the solution was less clear. In general working-class housing, too, was directed to the periphery, but not the suburban periphery. The new working-class districts were drab, sprawling, and socioeconomically segregated, and since they were built adjacent to extensive factory districts, they were subject to visual blight and air pollution. And in some American cities, there continued to be a shortage of housing.

The Committee on Community Planning (CCP) of the American Institute of Architects was formed to articulate an alternative to the way housing was planned and constructed for the working class. As reformers, the community planners hoped to see government play a broader role in subsidizing housing, and they visualized a whole new technique of planned housing. This technique included a new archi-

[3]See Jay Downer's "Bronx River Parkway," in *Proceedings of the Ninth National Conference on City Planning* (New York: D. C. McMurtie, 1917), pp. 91–95, which proposed to displace a number of poor inhabitants in the path of the parkway without provisions for resettlement. Beginning with an association of Fifth Avenue merchants in New York, business groups pushed for zoning legislation to protect their businesses from competing land uses. On the whole the application of zoning strengthened the functional differentiation of metropolis. On zoning see E. M. Bassett, *Zoning*, (New York: Russell Sage, 1936); R. Babcock, *The Zoning Game* (Madison: University of Wisconsin Press, 1966). On the City Beautiful and City Efficient: C. Tunnard and H. H. Reed, *American Skyline* (Boston: Houghton, Mifflin, 1953); C. Tunnard, *The Modern American City* (Princeton: Van Nostrand, 1968); Mel Scott, *American City Planning Since 1890* (Berkeley: University of California Press, 1969); Edward Relph, *The Modern Urban Landscape* (Baltimore: Johns Hopkins University Press, 1987). Jon Peterson, "The City Beautiful Movement," *Journal of Urban History* 2 (August 1976): 415–434, reveals the extent of upper-middle-class civic involvement in the City Beautiful Movement; and in Mark Luccarelli, "The Regional Planning Association of America" (Ph.D. dissertation, University of Iowa, 1985), pp. 29–31, I argue that the City Beautiful movement was an influence on City Efficient planning. Even though George Ford, a leader of the "scientific" planners, considered urban planning "99% technical," a science "as definite . . . as pure engineering," he made room for "aesthetic" and "social" concerns; see "The City Scientific," in *Proceedings of the Fifth National Conference on City Planning* (Boston: University Press, 1913).

tectural vocabulary adopted in part from the English garden cities, but it basic focus was the planning and construction of multiple housing units within the framework of an overall plan. This was the primary innovation of the CCP: linking housing and planning. In a later and quite different context, this linkage would be called planned unit development, or pud.[4] The goal of the CCP was not to build housing but to construct "communities" that would combine the best of the new techniques with the most desirable characteristics of an old urban neighborhood.

The importance of community as a social ideal for the CCP is apparent in the relation of its ideas to the theories and practice of social workers such as Mary Parker Follett, Jane Addams, and Robert Park, who looked to the local community as the locus of citizenship. The social workers' emphasis on neighborhood democracy and equality as ideals that would reshape the life of the urban working class was an influential formulation of radical progressivism. This ideal of *neighborhood as community* was an important model for the CCP, and eventually the RPAA would try to reach the social work community with its vision of planned new town development. The settlement house idea—working with the urban working class to strengthen citizenship by creating a public life for urban dwellers within their own neighborhoods—demanded the reconstruction of society at the local level and a movement against the massification and alienation of modern urban life. The social workers emphasized carefully managed structures, neighborhood associations, and educational activities as the basis for teaching democratic values and enabling civic participation. Later, the CCP would envision its built environment as a physical elaboration of these social and political structures.

The community ideal continued to be an essential idea of radical progressivism. But the opportunity to implement community planning came not as a result of the left progressives but in the wake of U.S. participation in World War I. As the nation mobilized for war, many reformers found it politically expedient to endorse the war effort and its nationalistic premises. Taking up Herbert Croly's call for a spirit of national unity, they sided with the state in its foreign

[4]See footnote 10.

adventure in exchange for the opportunity to press forward with a more active government at home.[5] Hence the unfortunate, perhaps ironic, alliance of "community planners" and other genuine reformers with militaristic nationalists—an analogue of cold war liberalism, the liberal "compromise" of the cold war era that exchanged support for the national security state for a mildly progressive domestic policy. Several RPAAers belong in this category: Stuart Chase served on the War Industries Board; Benton MacKaye wrote a Department of Labor report advocating the development of forestry villages based on the employment requirements of returning veterans; and most important, the community planners—including RPAAers Clarence Stein, Henry Wright, Frederick Ackerman, Robert Kohn, and John Irwin Bright—began the construction of the "war villages," the first federally financed housing projects.

The U.S. entrance into the war in 1917, three years after its commencement and at a point of crisis for the Allies, necessitated a rapid deployment not only of troops but of the economy. This was the first instance of economic warfare in the mature industrial era; the industrial resources at the command of the nation-state became the primary weapon. But to be effective, mobilization had to be rapid, and this required coordination of the economy. Given the constraints of time, coordination required extensive economic planning. The logic of necessity was compelling, and the federal government moved rapidly to implement the bureaucratic structures necessary to a planned economy. Federal administrators organized certain key sectors through such agencies as the War Industries Board, the Food Administration, the Fuel Administration, the Railroad Administration, and the Shipping Board. Production quotas were set, transportation prioritized, and organized labor accommodated. The logic of this economic restructuring also required an effort in the social realm (especially in the area of housing because a massive shortage was impeding full industrial production).

There was hesitation, however, about making such a commitment. After all, direct state intervention in the housing market might set a

[5]Croly's defense of American imperialism in Chapter 10 of *The Promise of American Life* (New York: Macmillan, 1910) is also instructive.

precedent for a permanent federal commitment to social welfare. And the prospect of unprecedented federal planning posed a problem of conscience for a political culture dominated by 19th-century prescriptions of laissez-faire ideology. Consequently, the most popular rationale for economic and (especially) social planning was its "emergency" character. From this point of view, federal economic planning was a *temporary* intrusion of government necessitated by the war.

By contrast, most of the reformers, including the community planners, seized upon federal wartime organization of industry as an opportunity to demonstrate the extent of social interdependency in a modern society and to use this as the moral justification for social planning. This certainly characterized the intent of Charles W. Harris and Frederick Ackerman, who undertook a series of articles in the *AIA Journal* to promote federal action on the housing front. According to Roy Lubove, these articles marshaled significant support for legislation enabling construction of federally financed housing for war industry workers.[6] Eventually 120 architects and planners were employed to build housing for approximately 180,000 workers and their families; the work was directed by two agencies, the division of Transportation and Housing of the Emergency Fleet Corporation (EFC) of the U.S. Shipping Board, and the U.S. Housing Corporation (USHC) of the Department of Labor. Yet the government's commitment to the housing program wavered because of the reluctance of many in Congress to intervene in matters that had always been left to the marketplace. There was also considerable disagreement among the housing experts about the appropriate nature and length of the federal commitment. Under the auspices of the National Housing Association (NHA), an important debate took place within the housing field between conservative and liberal reformers. Understanding their debate is essential to understanding the development of the community planning movement.

[6]See Roy Lubove, "Homes and 'A Few Well-Placed Fruit Trees,'" *Social Research* 27 (Winter 1960): 469–486; and *Community Planning in the 1920s* (Pittsburgh: University of Pittsburgh Press, 1963); see also Francesco Dal Co, "From Parks to the Region," in Giorgio Ciucci et al., *The America. City: From the Civil War to the New Deal*, trans. B. L. LaPenta (Cambridge: MIT Press, 1979).

In the *AIA Journal*, editor Charles H. Whitaker published the first of a series of articles entitled "What Is a House?" Whitaker argued that since war "brings nations face to face with national death," Americans might finally be compelled to examine their society. If so, they would find a nation that trails the entire "civilized world" in providing decent living conditions for its citizens, a nation that permits "one man bent on speculation which promises large profits to him" to take advantage of the "helpless community" by constructing inadequate housing. Intervention in the housing market, he concluded, is not only a "just government function" but an essential step toward national survival.[7]

To further his argument, Whitaker dispatched Ackerman to England to complete the "What Is a House?" series. The British had passed a Housing and Town Planning Act in 1909 and had begun to develop housing for workers. Ackerman was impressed by the decision to consider housing in relation to recreation, hospitals, churches, markets, and schools—to plan "complete communities." He suggested that the British example be applied in the United States: "the government should organize a separate department or non-profit government corporation . . . [empowered to] acquire land . . . plan new villages, install roads, sewer, water, and light, erect houses and other buildings of amenity required by these communities." Ackerman also challenged the architectural and housing establishment to recognize the implication of these measures: "it would be strange indeed, after this experience in what approximates a national syndication of production and collective provision which now holds sway [in England], if men should return . . . [to] life under the same and wasteful conditions which prevailed before the war." From the outset, therefore,

---

[7]Charles Whitaker, "What Is a House?" *AIA Journal* 5 (October 1917): 481–485; Whitaker's iconoclastic tendency led him to publish Louis Sullivan's "Autobiography of an Idea" in the *AIA Journal* over considerable opposition. The influence of Henry George on Whitaker was evident in his history of architecture: *Ramses to Rockefeller* (New York: Random House, 1934), which explained architectural development in terms of maximization of rent. Whitaker's interest in reform also led him to lend his good offices to the CCP and, later, the RPAA; some RPAA meetings were held at his New Jersey farmhouse.

Ackerman viewed the war preparations as an opportunity to push for a broader agenda directed at achieving an underlying national unity, an "integrated purpose" forged by planners working for social and economic solidarity. In calling for federal commitment to the housing area, he linked the planning and construction of "war villages" in England to the social evolution of industrialized society.[8]

Ackerman and Whitaker's proposal upset the status quo in the housing and planning fields by challenging the long-standing separation between "housers" and planners. Although the NCCP originally encompassed social workers and other reformers, the planners' new orientation toward transportation and zoning excluded other issues. In response the social reformers withdrew to create their own organizations. The NCCP became the forum for City Useful planning; and the most important housing group, the NHA, followed a conservative agenda centered on "restrictive" legislation regulating the construction of privately built low-income housing. This convenient division of labor went without serious challenge until Whitaker and Ackerman made their proposal.[9]

[8]Frederick Ackerman, "What Is a House?" *AIA Journal* 5 (December 1917): 591–622.

[9]According to J. L. Hancock, "Planners in the Changing American City, 1900–1940," *AIA Journal* 33 (September 1967), those interested primarily in housing issues separated from the NCCP in 1910. Their new organization, the NHA, led by Lawrence Veiller, pushed for the passage of building codes designed to regulate the height and increase the street frontage of residential buildings. Veiller had been instrumental in the passage of New York's famed tenement house laws (the most important of which was passed in 1901), which established requirements for lighting, bathroom facilities, and fire walls in new buildings. Veiller's approach to social problems reflected the crude environmentalism characterized in the early, privately financed philanthropic societies such as the New York Charity Organization. As historian Robert Bremner, *From the Depths* (New York: New York University Press, 1956), p. 130, explained, the social reformers thought of poverty as a "social disease as surely curable as tuberculosis, typhoid and yellow fever." Indeed the success of the public health movement in ridding the cities of the frequent epidemics that once had periodically threatened thousands of lives became a metaphor for the cleaning up of the social problems by the alteration of the physical environment of tenement dwellers. See also Roy Lubove, *The Progressives and the Slums* (Pittsburgh: Uni-

Ackerman had asked for federal action, but he found himself immersed in controversy. In early 1918, while Congress stalled on housing legislation, the NHA sponsored a symposium in Philadelphia on the issue of war housing; a three-way division of opinion emerged. Ackerman's ideas were well received by some planners, while others expressed the view that only temporary barracks would be necessary to meet the emergency. The leaders, Lawrence Veiller of the NHA and Frederick Law Olmsted Jr. of the NCCP, took a compromise position that called for the construction of permanent housing under temporary federal ownership. Thus, a year later, at the annual NCCP meeting, Olmsted insisted that since the war was over the government must sell the projects as soon as possible. With the "emergency" ended, Veiller and Olmsted agreed that the dominion of the private market in housing should be reestablished.[10] Barely tolerated during the wartime emergency, the experiment was terminated as soon as the war concluded.

---

versity of Pittsburgh Press, 1962). One noteworthy challenge to Veiller's agenda, and the inattention of planners to housing issues was registered by Benjamin Marsh, general secretary of the New York Committee on Congestion of Population which was instrumental in founding the NCCP. Marsh had been influenced by Henry George's proposal of land redistribution through the imposition of a single tax on land. In an *Introduction to City Planning* (New York: n.p., 1909), he suggested policies such as extension of mass transit lines to encourage urban diffusion; he was also one of the first Americans to favor building garden cities.

[10]On the first conference, see NHA, *A Symposium on War Housing* (New York: NHA, 1918): one of those in attendance, Philip Hiss, a New York architect and chairman of the Committee on Housing of the National Council of Defense, compared workers to "racehorses" who simply required "good stables." The compromise position was summarized by F. L. Olmsted Jr., "War Housing," in *Proceedings of the Tenth NCCP* (Boston: Taylor, 1918), pp. 97-103, who argued that government intervention was necessary only because the free market did not have time to respond to the crisis. Veiller, *Symposium*, pp. 120–121, argued, predictably, that the only "lesson" to be derived from the war housing program was that private housing projects could benefit from more inclusive building codes to regulate space, ventilation, fenestration, etc. Responding to a proposal to turn the projects into nonprofit cooperatives after the war, Veiller insisted on selling to companies or individuals and resorted to red-baiting: "the whole community owns

For most advocates of the federal program, including the CCP, the opportunity to plan "complete communities" that provided housing for working people had been long overdue. Imagine their frustration when they came to realize that the United States had not turned a corner in its history: there would be no permanent federal commitment to housing at that time. Ackerman's disappointment was acute; he realized that his attempt to build a consensus for subsidized housing had failed.[11]

Ackerman and Whitaker had created a synthesis of planning and housing that stood out from the proposals of earlier planners and housers. Community planning drew from two sources: radical progressivism, with its concern for engendering local, participatory civic

---

all the real estate. No individual is allowed to own a single bit. Where else could you have a more extreme form of communism than that?" One of the more thoughtful critiques of Veiller was advanced by E. E. Wood, *Housing of the Unskilled Wage Earner* (New York: Macmillan, 1919); she argued that Veiller stood for the "restrictive" or regulatory legislation of the past, while the future demanded "constructive" legislation, a more positive role for the state, to address the housing needs of those whose incomes preclude competing in the private market. Wood later became a member of the RPAA. But Olmsted in "Planning Residential Subdivisions," in *Proceedings of the Eleventh NCCP* (Boston: University Press, 1919), established what has remained the attitude of mainstream planners toward the techniques pioneered in the building of war villages. They could not become the basis of future development because this would constitute an unwarranted intrusion of government in the private market, yet they could inform the efforts of private builders. Ultimately the product would have to be judged acceptable in the marketplace. This conceptualization explains the subsequent history of planned unit development (pud) which in the United States was a tool that remained largely in private hands.

[11]Ackerman lashed out at the privileged position of the upper classes; as for city planners, they had shown that their "so-called 'fundamental principles'" were nothing but "suggestions of what is deemed politically expedient at the present juncture." See Frederick Ackerman, "Where Goes the City Planning Movement?" *AIA Journal* 7 and 8 (December 1919-October 1920): 591–622, 518–520, 284–287, 351–360. In "Our Philosophy of Restriction" and "Production via Curtailment," *AIA Journal* 9 (April and August 1921): 142–143, 264–265, Ackerman's reliance on Veblen's theories to explain the failure of a capitalist "money economy" to create full industrial productivity is evident.

structures; and social democracy with its emphasis on using state planning to overcome socioeconomic inequities. By making planning and housing speak to these social concerns, Ackerman and Whitaker helped to enlarge the meaning of social architecture. Though they failed to sustain interest in community planning after the war, the movement had already bequeathed two important legacies: a set of planning principles and techniques adapted largely from the English garden city idea, and a social democratic and communitarian political context for applying these techniques. Both were essential to the RPAA and to communitarian-progressive planning during the New Deal. The political framework informed Clarence Stein's attempt after the war to develop publicly supported community housing in New York State. The planning techniques shaped the modest private efforts of the RPAA-linked City Housing Corporation in the building of Sunnyside Gardens in Queens, New York City, and Radburn in Bergen County, New Jersey (see Chapter 8).

The most important contribution of the community planners was their understanding of the need for a social architecture. Because the planned villages were to be built at the periphery of existing cities, the entire physical layout of the community was planned. Planners could create a "whole environment." While this may be understood as an unwarranted enlargement of the planner's power, the intention was to provide a basis for an enlarged civic life. The emphasis was on the creation of public settings: common courts, green spaces, and "community centers." The idea of a commons—of a public place for civic life—became an important element of the idea and design of community. Second, they saw these planned communities as an answer to the imposing and inhumanely scaled development of urban places. *Scale* was the critical term; community planning intended to restore a human scale that had been lost in the process of creating megacities. And third, planned communities would not be for the middle class alone; they would provide a new way of life for the vast urban working class that in many cities had been forced to accept inadequate tenement housing.

Modernist in conception, community planning would adapt the model of mass production to human ends. Thus we might say that a unified process of planning and construction would *produce* community as a unified place; design would reflect the efficiencies of a unified production process and the community values of compact scale and social cohesion. Planned developments would be designed and constructed as a whole so that laying out infrastructure and housing as a single process reflected the technique and efficiencies of large-scale industrial production.[12] Community planning thus gathered a set of planning principles, including curvilinear streets that followed topographical features, street widths adapted to function (especially arterial versus residential), common courts, open green spaces, and neighborhood community centers. Taken together, and combined with architectural techniques such as the use of multifamily row housing and asymmetrical groupings, planning of this kind necessitated the reorganization of land division and street layout. This was intended to alter the conventional grid layout of American cities. The grid of streets that characterized American urban areas offered the advantage of flexibility and met the increasing demand for more intensive land use. Property along a typical street originally subdivided for the purpose of building single-family housing could easily be adapted twenty years later to commercial usage. This was in great part due to the maximization of street frontage in a grid plan. Community planners abhorred this arrangement because of its optimization of uncontrolled growth; in their eyes it permitted and abetted the continuing buildup of concentrated downtown areas, the uncontrolled dispersion of other urban functions (especially manufacturing), and the accompanying destruction of residential neighborhoods. While in retrospect we have come to appreciate the

[12]Principles of community planning are formally expressed in American Institute of Architects, *Report of the Committee on Community Planning*; reprinted in R. Lubove, ed., *The Urban Community*, (Englewood Cliffs, N.J.: Prentice-Hall, 1967), pp. 115–144. The social democratic implications of these ideas are most forcefully developed by Ackerman in "What Is a House?" and clearly reflected Veblen's influence. Ackerman's conceptions of planning were accompanied by his belief that the beautiful could be achieved in an architecture that reflected the process of mass construction and the requirements of a social setting. In this respect it is interesting to compare his point of view with that of the established academic architects; see Ralph Adams Cram and Frederick Ackerman, book reviews of *A Guildman's Interpretation of History*, by A. Pentry, *AIA Journal* 7 (October 1920): 373–376.

mixed-use neighborhood, the community planners understood that without planning, such neighborhoods were inevitably destroyed.

Negatively, community planning may be seen as an overly technical and nonpolitical response to the problems of social dislocation due to infrastructural modernization, contributing to what Francesco Dal Co considers the use of planning "as an instrument of order."[13] Positively, however, community planning permits the architect-planner to conceive of a project holistically, to consider not only street layout, open space requirements, and housing design but their interrelation. This is what was meant by "comprehensive" planning for the creation of "permanent communities." As a model of urban development, it accommodated growth not by the more intensive redevelopment of existing urban areas but by the building of additional communities. Writing for the *AIA Journal*, architect (and eventual RPAAer) John Irwin Bright understood these implications. His article on London demonstrates the connection between community planning and garden cities as a means of handling metropolitan growth; and it shows the influence of the London-based Garden City and Town Planning Association (see Figure 7).[14]

The social architecture of the CCP was an important influence on the RPAA. The CCP demonstrated that community planning can serve social ends, improving conditions of life for the working class while providing an environment—not unlike the urban neighborhood— conducive to civic life. Clarence Stein's partipation in the movement to preserve Chelsea, a fine old New York neighborhood, demonstrated the relation of this civic ideal of the neighborhood to garden city community planning. Familiar with the neighborhood through his involvement with the Ethical Culture Society's Hudson Guild settlement house, Stein feared that the City Planning Commission's "West Side Improvement Plan," which called for the New York Central Rail-

[13]See Francesco Dal Co, "From Parks to Region," in Ciucci, *American City*, p. 220.

[14]John Irwin Bright, "Unpopulating London," *AIA Journal* 8 (October 1920): 354–356.

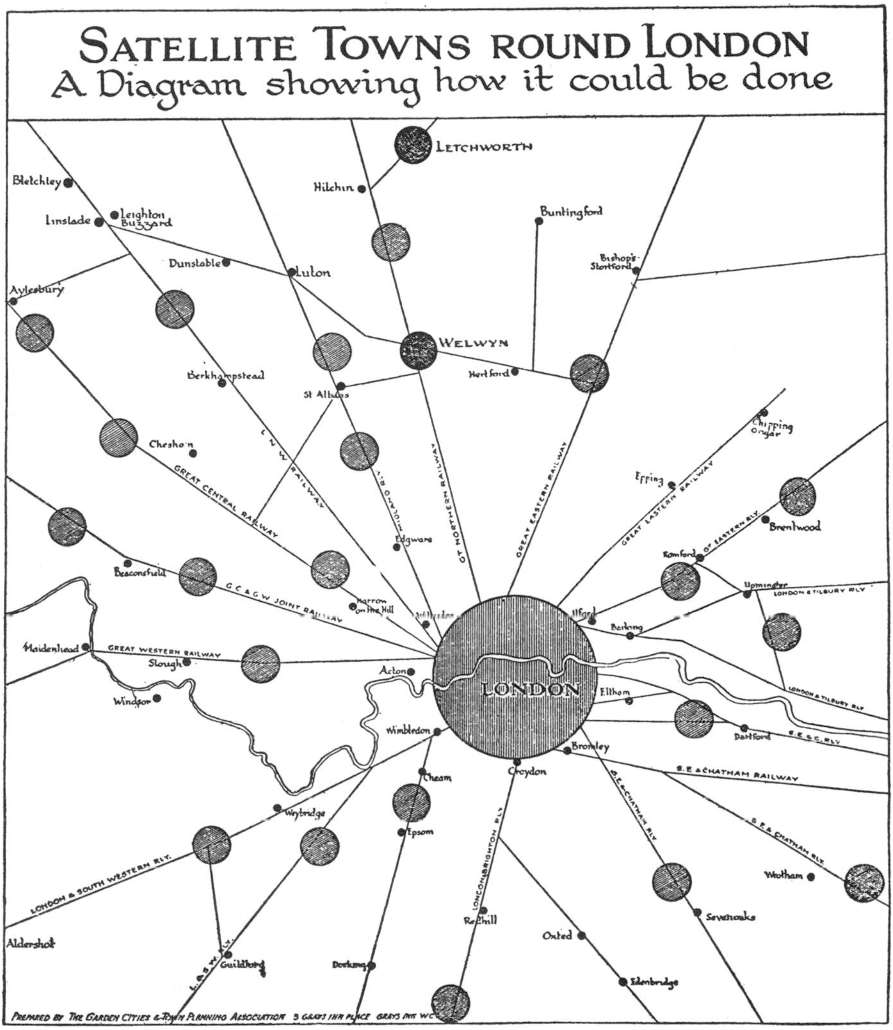

FIGURE 7. John Bright's 1920 article on controlling population growth in metropolitan London by use of satellite garden cities shows the concern with metropolitan regional planning. Bright was a member of the Committee on Community Planning of the AIA, a forerunner of the RPAA.

road to construct an elevated railway alongside Tenth Avenue, would deal a cataclysmic blow to Chelsea, a mixed-use neighborhood of houses, stores, and industrial shops. Chelsea, Stein thought, was an attractive neighborhood that maintained its identity in part through a centrally located park and local school. Building the rail line through the neighborhood would threaten its character by spurring further commercial and industrial development. Yet Stein strongly supported the City Efficient agenda of improved transportation linkages to rationalize the city by developing concentrated industrial districts. The "West Side Improvement Plan" was a perfect opportunity to encourage this kind of development.

In regard to the immediate problem, Stein negotiated a compromise. The rail line would be built; its impact would be mitigated by the further development of the Chelsea Park as a neighborhood civic center. As a member of New York's City Club, Stein was able to raise private monies to this end. Privately, however, Stein knew that his solution was not generally satisfactory; it was only the best that could be managed under the circumstances. And as a result of his experience in Chelsea, Stein determined to work toward creating conditions that would eliminate the need to make painful choices between residential neighborhoods and infrastructural and industrial modernization. He now placed his hope in the building of planned communities in peripheral areas. This would make possible the simultaneous fulfillment of the goals of community building and industrial modernization.[15]

❖

As Stein's reflections on Chelsea indicate, the RPAA took the position that the established cities could be saved by building garden cities in the countryside, but this belief also implied that the garden city could provide a much better way of life, contributing to the general turn against cities. In effect, Whitaker and Ackerman had proposed community planning as a means of simultaneously addressing social inequity, civic rehabilitation, and industrial productivity. By rationalizing the production of place, the community planners hoped to mediate between the social needs of the working population and

[15]Clarence Stein, "Neighborhood Planning in Chelsea," unpublished MS, 19 April 1917, Clarence S. Stein Papers, Olin Library, Cornell University, Ithaca, New York.

the economic needs of industry. This same idea motivated Stein's response to the Chelsea crisis: the needs of the community balanced with those of modernizing industry. And this was precisely the position that informed Stein's positioning of the RPAA: a "responsible" and "realistic" alternative to chaotic and uncontrolled growth.

He made his case for controlled growth in "Dinosaur Cities," published in the RPAA's special edition of the *Survey Graphic*. In his article Stein conjoined the progressive planning of the CCP with Mumford's regionalism. He argued that RPAA proposals for regionalization of the metropolitan city were not only reasonable but the best response to what he presented as an inevitable and on-going process of urban restructuring and modernization. Reviewing multiple issues, from housing and transportation to infrastructural and industrial growth, Stein pictured an urban system woefully inadequate to accommodate the technological and human demands attendant on this restructuring.

Stein rooted the urban crisis in the accelerating demand for facilities that existing urban centers could not provide. Housing was a case in point:

> Superficial observers talk of this housing breakdown as if it were a product of the [First World] war. On the contrary, there is a chronic deficiency that has been piling up in every great city—in London, Paris, and Berlin, as well as in American cities—for the last hundred years. In the great city there are not enough decent quarters to go around; and even the decent quarters are not good enough.[16]

For Stein, the housing shortage was symptomatic of the reality that lay behind the dream of the "great city"—New York and all its lesser imitators. The great metropolis, the center of culture and urbanity, has little to offer the mass of its citizens: "What part does art, literature, culture or financial opportunity play in the lives of the millions of men and women who go through the daily routine of life in our great urban districts?" (*DC*, p. 66). Not only is the idea of the great

[16]Clarence S. Stein, "Dinosaur Cities" *Survey Graphic* 7 (May 1925): 134–138; reprinted in Carl Sussman, ed., *Planning the Fourth Migration* (Cambridge: MIT Press, 1976), pp. 65–74, quotation on p. 67; hereafter referred to as *DC*.

cultural center a sham for the masses who live there, but even the "prospects for decent human living" as measured by the availability of housing are very dim.

Stein observed that since the industrial revolution the city had become the victim of its own success. Its aim was to grow, but that very growth caused a series of economic and environmental "breakdowns" that threatened its survival. The deterioration began with severe housing shortages but soon included the water and sewer systems and then the transportation system. The growth of the great metropolis made it unworkable; the diverse balance of places and human activities that made urban life possible was threatened. For example, the "spread of sewerage in the Hudson, the East River, and the Harbor has not merely destroyed the opportunities for bathing and caused the practical disappearance of the North River Shad . . . [it] also . . . cut the city off from 80 percent of its shellfish, increased the dangers of typhoid . . . and now threatens the bathing beaches of Coney Island and Brighton" (DC, p. 68). Stein realized that the ecology of the city was being destroyed: the city and its immediate environs were losing their diversity, their capacity to support a healthy variety of human and natural activities. He understood that the city is a habitat, one that encompasses wild organisms (he mentioned fish and shellfish but could have added birds and other residents of the marshes that surround the city) and human beings.

The response to such "breakdowns" has inevitably been technological. Technological innovations are applied not in the interest of restoring urban diversity but in order to sustain growth. Thus the pollution of the rivers is addressed by building new sewage treatment systems that can overcome at least the worst effects of the problem. But the reliance on large-scale plants and equipment to treat sewage is expensive; it does not address the loss of habitat and adds further costs to the already high price of living and doing business in an urban area. This is even clearer in respect to transportation.

The growth of the city has always been marked by congestion—both the lack of space and the slowing down of traffic. The elevator was an early technological "miracle" that addressed the need for more space in the compact center city areas. Elevators made it possible to overcome the shortage of space in the city center by permitting the construction of taller and taller skyscrapers. This enabled builders to concentrate larger and larger megastructures into small areas of the city. The increase in urban densities has the effect of creating *"from two to six cities . . .* piled up one above the other" (*DC*, p. 69).

The creation of additional urban vertical space entailed the peripheral extension of the city. The response to the congestion of the center city is the spread of residential districts further and further from the core. Sprawling residential districts are required to house the populations that each day find their way into the city center. This mass movement of population requires the introduction of additional technological innovations in transportation:

> As the city increases in height it increases in area; for the railroad and subway must be introduced to carry the main load of passengers from the central district of skyscraper offices and lofts to the outlying areas. When the vacant land on the outskirts is filled up, the net result is congestion at both ends. This causes a demand for additional means of transportation. Beyond a point which big cities reach at a very early point in their career, more transportation routes mean more congestion. (*DC*, p. 70)

The system depends on commutation, and the growth of the city entails even more elaborate commutation. But spreading the population into a large area does not relieve congestion; it simply disperses the population. In fact, Stein argued, the more we attempt to solve congestion by the application of new transportation technologies, the greater the congestion of the urban center will become.

Today we have experienced the spread of relatively low-density suburban development and now regard the congestion, the high densities, of the inner city as not all bad. These densities are what make an urban culture possible, because the greater diversity of a large population sustains an interesting multiplicity of activities. But at the time Stein was correct in raising questions about the impact of ever increasing concentrations of population. His objections to this process of accelerating centralization were social and economic. He called attention to the city's existence as an economic entity that relies on street transportation for the movement of supplies and products. He argued that this street system of transporting goods and population was breaking down under the load, worsening the congestion of traffic, impeding the operation of industries that rely on the streets for transportation of goods. This lessens the probability that industries will be able to modernize and increase their productivity. Efficiency in transportation is necessary to industrial restructuring and expansion, all part of "the normal processes of industry" (*DC*, p. 71).

Stein was also aware that intensive development of the city, especially the construction of megastructures, erodes the amenities of urban life. The city's core areas become functionally differentiated; specialized zones for business and commerce deteriorate the city, making it a less desirable place to live. And as a good regionalist, he noted that as the city spreads outward, it consumes the adjacent open lands. Stein was appalled by the prospect of sacrificing entire urban neighborhoods to open space for large factory complexes or permitting industrial diffusion to close off the remaining open land around the great cities. The attempt to redevelop existing urban neighborhoods (such as Chelsea) by improving transportation and building larger factories would destroy urban life. Yet he predicted the sacrifice of entire urban neighborhoods would fail to meet industry's requirements for space. Even if entire neighborhoods were razed to make room for industrial and commercial growth, the life of the great city, the concentration of activities—the little factories, docks, offices, residences—would create a bottleneck in transportation of goods that would compel decentralization. What sense did it make to continue to have central urban areas as the centers of production when the lack of space forced population growth toward the periphery? What sense did it make to build an even more centralized system of transportation if some portion of the city's population and industry could be accomodated in new garden cities?

The development of mechanical transport (railroads, subways, and elevators) had done its work too well. Having created the city as an engine of rapid industrial and commercial growth, the new technologies were now threatening its survival. The great city itself was in part the creation of the 18th-century revolution in the technologies of energy production and transportation. Steam engines required the delivery of coal, and the railroads that supplied it created a centralized distribution system that drew raw materials from the hinterlands to a few central locations; this resulted in centralized production that made possible rapid urban growth. In effect the railroad and the steam engine made the metropolis possible, and new technologies of the early 20th century, the subway and the elevator, permitted its continued growth by making it possible to (temporarily) address the ecological breakdowns. But fixing the breakdowns simply whetted the appetite for more growth: more systematic transport increased the specialization of functions within the urban zones and the use of more and more land surrounding the city. These changes undermined the ecology, both natural and human, of the city.

In this sense, said Stein, the great city is a "dinosaur." The city is too densely populated, too congested, too dependent on distant sources of water and sewage disposal, too much at the mercy of elaborate systems of transportation, and, crucially, too expensive, both economically and environmentally. It is also too fragile to withstand the impact of modernization. Its fragility is inherent in the ecological balance of urban life, a balance of places, peoples, and nature that took centuries to create. Stein understood the danger in this situation. The demand for amenities, not only housing but areas suitable for industrial and commercial redevelopment, would eventually undermine what remained of livable areas within the historic urban cores. The crush of population and industry would undo what remained of the historic city.

The point, therefore, is *not* that Stein looked forward to the death of the great urban centers. The great city must survive, but it could no longer be the engine of growth and development. It must spin off population and industry, creating not "satellites" but new urban places with their own mix of urban functions. Only planning informed by a regionalist perspective could make this possible. Stein thought that planned decentralization was necessary to bring about an historic change in urban form: the end of the centralized metropolis—the sprawling city with its vast residential districts, factory zones, and fashionable suburbs all focused on a central business district. Garden city regional planning could make possible a decentralized metropolis: a metropolis that would supplement the existing urban cores.

In the new context created by the technologies of decentralization, Stein considered the emphasis on transportation a "technological fix" that substituted increased mobility for intelligent planning. The new technologies (electric transmission lines and automobiles) made possible a new era of decentralized life; these technologies proposed a different urban form: the garden cities of a regionalized metropolis. In effect this form would use new technology to lessen the demand for transportation.

The issue, Stein somewhat mistakenly argued, is between building onto the city as it is or building outside of it. If the former is chosen, a formless, inefficient, and environmentally deficient megacity will be created; if the latter, new forms appropriate to decentralization will emerge:

> We are still in the day of postponement; but the day of reckoning will come; and it behooves us to anticipate it. The question then will be whether industries [and population] are to migrate into the free area that lies immediately around the great city, or whether it will not pay, once moving must be faced, to locate at some point in a much larger area, where land prices are not so high and where a finer environment may be provided more easily, without the risk of being gobbled up eventually. (*DC*, p. 72)

The specter of land being "gobbled up" is telling. The language he used suggests the extent to which Stein's conception of planning was informed by the regionalist concern for preservation of natural lands. Yet in his discussion of land use and the fate of the city, Stein was both right and wrong. He was right that we faced (and continue to face) a choice between planning on the one hand and wasting rural and natural landscapes on the other. He understood, therefore, that the direction of growth (for some time) would likely be toward the periphery and that the choices we faced centered on how to shape that peripheral growth. Yet the choice was not exactly as he envisioned it: the sprawl that we find ourselves afflicted with is not the sprawl of the old centralized metropolis in which activities continue to be channeled to the urban core but the spread of the new (postwar) *diffused metropolis*. This suggests the extent to which Stein, the RPAA, and everyone else failed to consider that decentralization might gain such momentum that it could threaten the economic viability of the center cities.

In this respect Stein did not distinguish himself from other commentators on the future of the city. Like others, he simply assumed that regardless of the degree to which we decentralize, "the city would always be there."[17] He was unaware of the threat that decentralization posed to the existing urban cores. The new technologies that Stein thought critical were used not so much to create new forms (regional cities) as to destroy the old form (the centralized metropolis).

Still, we can see that to some extent the decay of the urban core was the result of the particular kind of decentralization we have experienced. When decentralization proceeded (aided, as we shall see, by new federal policies), it was devoid of the kind of regional planning that Stein and the RPAA thought necessary to the process. This produced a diffused metropolis—a sprawling, visually blighted, and functionally differentiated amalgam of existing towns and cities, residential subdivisions, industrial and office "parks," shopping centers, and malls. Here growth not only consumes the surrounding countryside, it deprives the urban core of its integrity, destroys urban amenities. Diffusion must be contrasted with the *planned* decentralization that Stein and the RPAA advocated: planning shapes the "outflow" from urban areas into a pattern determined by regional geography, enhanced by regional landscape and structured by a coherent and balanced community life.

The garden city or new town is the instrument of this process of decentralization. Designed to avoid the conflict between the need for more space and existing urban integrity, the new town could meet the requirements of modernizing industry and the human requirement for affordable housing and integrated communities. The new town also balanced a program for modernization of infrastructure, with systematic land use planning that could permit the creation of decentralized urban areas, each far more self-contained than existing sections of the metropolitan complex. Only the planned community would allow the relative proximity of workplace and

---

[17]According to Robert Fishman, *Bourgeois Utopias* (New York: Basic Books, 1987), p. 192, most planners well into the 1950s failed to recognize the impact of highway construction in undermining central business districts.

residence. It is true that as conceived by the RPAA, the garden city would be spatially segregated, such that residential areas could be separated from industry and, to a lesser degree, from commerce.[18] But in large part this reflected the attempt by planners to accommodate the land usage and transportation needs of the large smokestack industries that dominated economic production at that time. The idea of mixed land use as essential to an urban experience makes no sense if urban residents find themselves adjacent to huge industrial complexes.[19]

Even with its limitations, Stein's case for replacing the "dinosaur city" with a new kind of regionalized metropolis was one of the most compelling put forth by the RPAA, in large part because of the way he positioned it. Without a progressive popular constituency as an important catalyst to change, Stein knew he had to tailor his message to an elite audience: business interests, professional elites, public servants, and social reformers. For them, he proclaimed the possibility of achieving a new consensus—a move not dissimilar to the direction that American liberalism took in the 1930s. It is necessary, Stein argued, to construct out of the built environment a conception of public interest that has a corresponding benefit for business interests. And the relationship among social issues, transportation, the urban structure, and the natural environment was the logical place to work out this new consensus. Therefore he did not question the need for technological modernization or its consequences; for Stein, industrial diffusion was the inescapable consequence of adapting land use to the requirements of modernizing mass production industries. The scene of action, he insisted, was shifting away from the center city to the periphery, for it is only at the periphery that the

[18]In *Communitas* (New York: Vintage, 1960; orig. pub. 1948), esp. pp. 25–27, 32–35, Paul and Percival Goodman argue that in its differentiation of space, the "garden city" isolated life from work, reinforcing the tendency of industrial cultures to translate living into an assemblage of amenities.

[19]On this see Mumford's text and photographs in *C of C*, pp. 436–437.

population increases and modernized, "neotechnical" mass production industries could be accommodated. Stein's argument was that the RPAA's garden city regional planning recognized these requirements and offered a viable and reasonable alternative to business as usual.

Stein's positioning of the RPAA was affirmed by Stuart Chase's work. Contending that we need a rational economic system, Chase looked to a planning agenda capable of eliminating "the waste which flows from current transportation methods" and "which arises because communities are not regionally planned." In "Coals to Newcastle," his contribution to the *Survey Graphic's* regional planning number, he drew on both the environmental model of conservation regarding "waste" of natural resources and the social lessons Progressivism learned from unregulated competition in the Gilded Age.[20]

Chase contextualized "efficiency" in terms of industrial location: relocation provides an opportunity for modernization of plants and

[20]Stuart Chase, "Coals to Newcastle," *Survey Graphic* 7 (May 1925): 143–146; reprinted in Sussman, *Planning the Fourth Migration*, pp. 80–88, quotes on p. 82. Chase began his career as an economist for the Federal Trade Commission; during World War I the Harding administration dismissed him because he objected too vehemently to business war profiteering. In 1921 Chase joined the Technical Alliance, a group of radical engineers influenced by Veblen; with George Soule he also cofounded the Labor Bureau, which produced a monthly newsletter, *Facts for Workers*. Within a few years Chase published his most important book, *The Tragedy of Waste* (New York: Macmillan, 1925), in which he popularized and adapted Veblen's critique of capitalism to the new situation occasioned by the boom of the 1920s. In place of Veblen's analysis of "underproduction," Chase saw "waste" as the fundamental characteristic of "business enterprise." The misuse of resources and the overproduction of luxuries meant that "full" production did not bring prosperity for all. He prophesied the replacement of a chaotic, "wasteful" system of production and an "acquisitive" culture with a "functional society" planned by technocrats. Consequently, he extolled the virtues of Soviet-style planning. See also Stuart Chase, "Industry and the Gosplan," in S. Chase, R. Duncan, and R. G. Tugwell, eds., *Soviet Russia in the Second Decade: A Joint Survey by the Technical Staff of the First American Trade Delegation* (New York: John Day, 1928), pp. 14-54; and Robert B. Westbrook, "Tribune of the Technostructure," *American Quarterly* 34 (Fall 1980): 389–399.

equipment. He cited the successful operation of a Ford automobile plant to prove that modern "highly standardized, highly subdivided industry" can successfully relocate, even to rural areas. In fact, as Chase was well aware, business was anxious to disperse its plants from urban areas, where costs were increasing and the urban infrastructure limited the potential to expand and modernize facilities. (At the same time business was anxious to escape the union domination of urban areas, an issue Chase and the RPAA did not discuss.) In keeping with the RPAA's theme, Chase contrasted *diffusion* of industry into areas adjacent to metropolis to an orderly *decentralization*. He looked upon the ongoing second industrial revolution as an opportunity not only to revamp the system of production (as industries were doing) but to consider the location of industry, its utilization of natural resources, and the levels of production. According to Chase, the regionalist principle for planned economy was simple: factories should be built as close as possible to the places where raw materials were extracted. Chase extolled this as a great "efficiency" in production that would eliminate the cost of transporting raw materials to centers of production. It would be less expensive to transport finished goods to urban markets. Second, production levels must be planned. The new regionally oriented economy could be saved from the spasms of over- and underproduction by government regulation. In effect, Chase's proposals extended the model of private ownership and state planning and regulation first experienced in the United States during World War I.

CHAPTER 8

Planned Decentralization: The Road Not Taken

Given its divergence from planning orthodoxy, some historians have pictured the RPAA as working in "intellectual isolation" from mainstream planners; the truth is that the group was anxious to engage other planners.[1] Stein and Wright attended conferences and lectured at universities such as Cornell, Illinois, and Wisconsin, and they invited planners to attend RPAA meetings.[2] Under Stein's leadership the RPAA embarked on a three-part strategy for marketing its ideas: (1) winning planners over to the RPAA's version of regional plan-

---

[1]Roy Lubove, *Community Planning in the 1920's* (Pittsburgh: University of Pittsburgh Press, 1963), p. 1, says that in making a "sharp break" with the planning mainstream, the RPAA was "in full rebellion against metropolitan centralization and suburban diffusion alike." In his introduction to *Planning the Fourth Migration* (Cambridge: MIT Press, 1976), p. 43, Carl Sussman characterized the RPAA as having spent years in "intellectual isolation."

[2]Henry Wright to Russell Black, 2 January 1929, Black Papers, Olin Library, Cornell University, Ithaca, New York, describes a series of planned lectures.

ning, (2) building private-public partnerships for garden city housing in New York State, and (3) constructing model residential projects.
One of these goals, to identify a common ground with mainstream
planning, was realized: Stein and Wright's tireless efforts to promote
their agenda for regional planning and new town development received a generally favorable hearing from planners.

The single most important aspect of this effort, however, was Stein's
political work in New York State, and this did not meet with success.
Over a ten-year period, Stein cultivated civic and business leaders,
hoping to establish a consensus for metropolitan decentralization
and state-subsidized housing. He attempted to work out a private-
public partnership to build new communities and to institute regional design planning. This ultimately led to the formation of the
New York State Commission of Housing and Regional Planning.
Under Stein's chairmanship, the commission became a voice for comprehensive planning. Only regional planning, the *CHRP Report* of
1926 asserted, could save us from becoming "passive creatures of
circumstance . . . [rather than] the creators of our future."[3] Good
regional design requires "technical insight" and "social vision." In
the absence of this kind of planning, we could expect uncontrolled
diffusion of industry, commerce, and population.

At the time the *CHRP Report* was written, interest in regional planning was growing. Its currency was indicated by the NCCP's decision in 1925 to include "Region" in the name of its proceedings:
Planning Problems of Town, City and Region. But "regional" planning
turned out to be what was commonly called the "highway-and-parks"
approach. This meant planning for a system of metropolitan highways. Many proposals linked highway development to the planning
of a system of parklands encircling the urban core; other plans were
more concerned with additional features of infrastructure. For instance, the director of the Los Angeles County Regional Planning

[3]State of New York, *Report of the Commission of Housing and Regional Planning to Governor Alfred E. Smith* (Albany: J. B. Lyon, 1926); reprinted in
Sussman, *Planning the Fourth Migration*, pp. 145–194, quotation on p. 194;
subsequent references will be to the *CHRP* and will refer to Sussman's page
numbers.

Commission emphasized coordination of flood control and sewage disposal efforts for his metropolitan area.

There are two important points about the highway-and-parks approach to regional planning. First, it was an outgrowth of City Useful planning (see Chapter 7) that emphasized the provision of infrastructure over aesthetic or social criteria. The provision of parklands was usually an afterthought or at best conceived of as alleviation of the visual boredom created by contiguous development. Second, this form of regional planning demonstrated little concern with what Mumford called the "natural region." Planners were responding to what they saw as the inevitable infrastructural expansion of the metropolis. Thomas Adams, one of the leading professional planners of the era, defined a region as a central city and its "contributory area," consisting of all adjacent communities "which are *daily dependent* on the central city for their livelihood, social life or supply of commodities." The purpose of regional planning was, as Adams put it, to accommodate the "normal and desirable *growth of the city into the surrounding area.*"[4] Adams's notion of the "urban region" conformed to the 19th-century centralized metropolis in which the center city absorbed the focus of "daily" activities while suburbs became the locus of residential development.

By contrast, the RPAA had worked to redefine the region on a the basis of geography: the region was a "natural" phenomenon. MacKaye believed in planning according to nature's "design." He proceeded from the wild to the rural to the urban: land use designation was

[4]Thomas Adams, "Regional and Metropolitan Planning," in *Proceedings of the Fifteenth NCCP* (Baltimore: Norman, Remington, 1923), pp. 1–32, emphasis added; other important planners included George Ford and John Nolen. Also see, G. Gordon Whitnall, "City and Regional Planning in Los Angeles," in *Proceedings of the Sixteenth NCCP* (Baltimore: Norman, Remington, 1924), pp. 105–110; Mel Scott, *American City Planning Since 1890* (Berkeley: University of California Press, 1969), p. 212; and Alan Altshuler, *The City Planning Process* (Ithaca: Cornell University Press, 1965), pp. 299–305.

fundamental to this approach, and new "regional cities" were to be laid out in conscious relation to wild and rural landscapes. We can see this technique in the *CHRP Report*. Even though it had to work with the somewhat arbitrary political boundaries of New York State, in some respects the report reflected MacKaye's concern for the conservation of rural and wildlands (see Figure 8). It emphasized the establishment of forest reserves and agricultural lands—features of a balanced "natural region"—that establish a natural framework for the "belts" of land suitable for decentralized regional cities.

Unfortunately there was a very limited audience for this version of decentralization. The era of the megametropolis, or "megalopolis," was at hand; growth was becoming increasingly concentrated in a few metropolitan areas. Most planners saw this as a natural development or as the consequence of an economic structure that could not be altered. Consequently, they did not concern themselves with

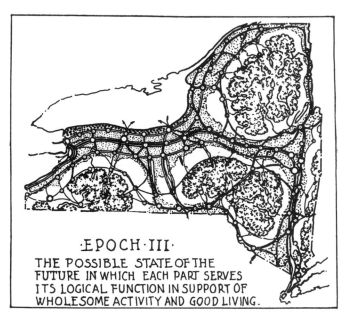

·EPOCH·III·

THE POSSIBLE STATE OF THE FUTURE IN WHICH EACH PART SERVES ITS LOGICAL FUNCTION IN SUPPORT OF WHOLESOME ACTIVITY AND GOOD LIVING.

FIGURE 8. Mumford and MacKaye's concept of urban development within a framework of natural features of the land found fruition in the *Report of the New York State Commission of Housing and Regional Planning of the New York*, chaired by Clarence Stein.

planning regional land uses or putting urban areas in relation to rural and wild landscapes; the entire emphasis of mainstream regional planning was with ordering the metropolis. If Stein wanted to find a common ground, he would have to direct RPAA ideas toward the issue of metropolitan decentralization. In effect this meant stressing the fundamental importance of the garden city as a tool for metropolitan decentralization, one of the major considerations of Stein's "Dinosaur Cities" piece.[5] The NCCP-sponsored International City and Regional Planning Conference held in New York in April 1925 took the garden city as a means of metropolitan decentralization as the central subject. This must have reinforced Stein's interest in the topic. Following the conference Stein arranged an informal reception for many of the European and British planners.

Defining the practical and theoretical contributions of garden city planning to the purpose of metropolitan decentralization had been the life's work of Raymond Unwin. Unwin, an important participant in the International City Conference and an associate of the London-based Garden Cities and Town Planning Association (with which the RPAA was officially affiliated), was the chief architect and planner of Letchworth, the first English garden city. Elaborating on Ebenezer Howard's original idea of building garden cities to relieve the population pressure on London, Unwin focused on the decentralization of the expanding metropolitan region: garden cities, he thought, should be considered as a particular, and superior, synthesis of existing conceptions of industrial and residential suburbs. The industrial suburb was a feature of continental European planning, and the residential suburb, of course, was a well-known feature of Anglo-American urban development. In the context of the expanding metropolis, the garden suburb would effect "a better distribution of population and a carefully planned and regulated development of our lands, which will check excessive concentration, and promote the localisation of life around more numerous centers." Thus, Unwin preserved something of the ideal of self-contained cities while plan-

[5]In effect this undid Mumford's attempt to make cultural regionalism the central concern of the RPAA by dropping "Garden City" from the group's name. See Chapter 5.

ning them as "satellites" of the urban core. A satellite should not be confused with a residential suburb: a satellite includes industry and business. It benefits from its close transportation linkages to the metropolitan core, but maintains an independence befitting a city. This would find fruition in the British New Town policy, instituted after World War II.[6]

The interest of Stein and the RPAA in Unwin's idea of metropolitan decentralization becomes apparent on reading the minutes of RPAA meetings. After 1927 the meetings were often attended by Philadelphia planner Russell Van Nest Black. Although he never became a full-fledged member of the RPAA (he was invited but never joined), Black was a major participant in the group's discussions. And like Unwin, Black argued that garden cities should be conceived as urban satellites, capable of absorbing the growth of metropolis. The garden city, or new town, would be connected by a network of highways to existing metropolitan cores.[7] Black's plan for the Regional Planning Federation of the Philadelphia Tri-State District called for the establishment of ten to twelve garden suburbs and a number of satellite cities to be separated by greenbelts from the center city and its ring of suburbs.[8] Despite the decentralization of population in satellite cities, Black assumed the continued growth of the center city. Had Black's plan been implemented, metropolitan Philadelphia might have resembled modern Stockholm: a network of spatially differentiated subcenters that would accommodate and coordinate

[6]Raymond Unwin, "Methods of Decentralization," in *International City and Regional Planning Conference* (Baltimore: Norman, Remington, 1925), pp. 152–155. Unwin's work clearly had an impact on Patrick Ambercrombie's influential 1943 plan for directing growth in the London metropolitan area.

[7]Stein had previously defined a satellite city "as dependent on its relation to a metropolis" and had juxtaposed that to a regional city, which is "primarily dependent on its relation to the natural opportunities of a region"; see RPAA, Minutes of Meetings, 1923-1933, meeting of 3 October 1927. For an account of Black's ideas, see RPAA, meeting of 8–9 October 1927; for Black's invitation to membership, see RPAA, meeting of 14 November 1931; all in Clarence S. Stein Papers, Olin Library, Cornell University, Ithaca, New York.

[8]The Plan is discussed in Scott, *American City Planning*, pp. 216–220.

residential, commercial, and industrial development, each subcenter remaining a part of the metropolitan whole. Interestingly, Black's conception of the garden city as satellite showed little consideration of the "natural region"; in this respect it was not much of an advance from John Irwin Bright's 1920 mechanistic notion of the use of garden cities for controlling metropolitan growth (see Figure 9).

Despite its limitations, Black's version of metropolitan planning must have been very appealing to those RPAAers anxious to establish liaisons with mainstream planners. And his plan for the Philadelphia metropolitan region seemed to confirm the importance the RPAA attached to geographically distinct urban centers, while recognizing the economic imperatives for larger and more integrated urban markets. We can see Stein and Wright's work for the CHC in the late 1920s (especially the design of the garden suburb of Radburn)

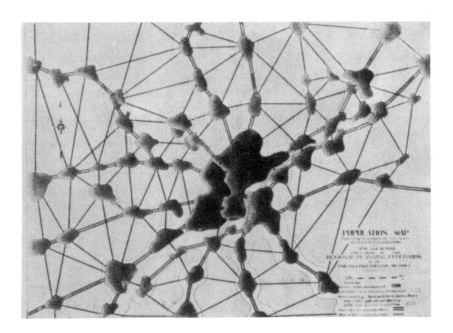

FIGURE 9. Russell Van Nest Black's plan for metroplitan Philadelphia indicates the popularity—and limits—of using satellite cities to channel urban growth. A good friend of Stein's and Wright's, Black nonetheless had little concern for the civic and ecological issues the RPAA raised.

in the context of both Unwin's and Black's ideas about metropolitan planning.

As a proposal for regional development of large urban areas, Black's plan suggests a different kind of metropolis, one more heterogeneous and less centralized than the metropolis of the RPAA's day: a diversified metropolis, an urban region composed of garden cities along with existing urban neighborhoods. From Mumford's perspective, this limited conception of metropolitan regional planning, though without the cultural and civic vision of ecological regionalism, had the potential to control the haphazard spread of the city across natural landscapes, and it might create a workable urban system on which to build a civic life.

In the United States, planning, whether conceived as regional or metropolitan proved to be a difficult task. For unlike in the Netherlands or Germany or virtually any other European country, in the United States land use decisions are left to the discretion of municipal governments. This practice of home rule is buttressed by the dependence of municipal governments on local real estate taxes. The system has contributed to the fragmentation of metropolitan areas because many different municipalities separately carry out decisions that affect the shape of the whole region. While in theory this is more democratic and opens the possibility for creating local "publics," in practice the U.S. federal system has turned over land use planning to local elites. These provincial upper classes have in turn assured that a highly profitable real estate industry dominates development decisions. The RPAAers understood the result: a destructive system of real estate speculation as the engine driving land use decisions. Frederick Ackerman put it succinctly: we have "woven [ourselves] into a complicated web of inflated values and capitalizations *which involve the necessity of [sprawling] growth and [urban] concentration.*"[9]

[9]Frederick Ackerman, "Our Stake in Congestion," *Survey Graphic* 7 (May 1925): 141–142; reprinted in Sussman, *Planning the Fourth Migration*, pp. 75–79, quotation on p. 77.

In responding to this situation, Stein was the consummate politician. Recognizing exactly what was at stake for the future of American cities, he carefully wooed planners and decisionmakers. His approach is apparent in the *CHRP Report*, which invited private interests and state bureaucracies to reach agreements, official and otherwise, to structure urban growth. The report addressed many of the players: "the Park Council, the Water Power Commission, the Conservation Commission, the framers of tax policy, the Public Service Commission, the highway authorities, manufacturers, bankers, railroads, power companies, local governments." It asked for an agreement to "coordinate their activities because their decisions are related . . . [and] interdependent" (*CHRP*, pp. 193–194). And in order to invite participation in building a *consensus* for change in land use practices, the *Report* left as open questions both the "multitude of details" of a regional plan and the actual means by which the planning would be carried out. Wright delegated to each "local region" the authority to "initiate action and carry out concrete details." He repeatedly reassured his audience that a "State plan will not attempt to limit this local action with hard-and-fast outlines" (*CHRP*, p. 193). "A plan like this," Wright reiterated, "is but a broad outline, simple in concept but complicated by a multitude of details and difficulties" (*CHRP*, p. 190). As "regional units" Wright had in mind intergovernmental agencies such as the Capitol (Albany) District Regional Planning Association, the Central Hudson Valley Regional Planning Association, and, most importantly, the Committee on the Plan of New York and Its Environs. These agencies had been "created through . . . cooperation" and foreshadowed the "creation of official planning boards" (*CHRP*, p. 182).

Stein masterminded this accommodationist strategy, and in his practice he reached out not only to municipal officials, but to private developers, arguing that state guidance of the development process would not end profitable real estate development. But when Stein attempted to engineer an aggressive state program for garden city housing development, his policies were defeated by those very interests he had attempted to accommodate. Twice in the postwar period (1919–1920 and 1923–1926), Stein had been appointed to state positions by Democratic governor Al Smith. With his associate Robert Kohn, Stein attempted a corporatist approach to New York's chronic housing shortage. Arguing that the rent control policies adopted by the state legislature would be ineffectual, Stein proposed building low-cost, for-profit housing with state assistance. Stein's plans called for the establishment of a state planning office empowered to authorize cities to acquire land and, in conjunction with local planning boards, to approve construction of garden city housing projects by authorized private contractors.[10] He suggested that local planning boards, nonpartisan agencies composed of business and civic leaders, oversee the projects. The local boards would ensure the involvement of community elites, and their public-spirited nature would guard against political corruption. In effect Stein was hoping to build on the savings-and-loan formula of directing local capital into housing: capitalism is mixed with civic obligation to provide benefits to the community.

Stein's formulation of a community-based development strategy is very suggestive of contemporary efforts to accomplish socially useful tasks without turning over control to the state. But rather than using the community focus to challenge the structure of power, Stein sought to co-opt it. True to Progressive Era politics, Stein wanted to keep the community planning effort above politics. The local boards would function as consensus builders to enable the municipality to assemble large tracts for new town development. With the approval of a state planning board and with monies provided by a state land bank, private developers could then carry out the development and sell the housing. According to Stein, the economy gained by the large scale of the development: the provision of low-cost state credit and the elimination of profits on land speculation would substantially lower housing costs, making the housing affordable to a wider spectrum of the population. State planning and the assembly of large tracts would prevent the wasteful misuse of land, while competent and sufficiently capitalized private developers could earn fair profits. The result would be affordable housing built as a complete community that would provide a new way of life for thousands of city dwellers.

[10] I call this approach "corporatist" because it amounted to a formalized political deal in which social reformers, political entrepreneurs, and large real estate interests all stood to benefit.

These proposals were not enacted, despite the governor's support and a determined campaign by Kohn and Stein to solicit the support of business, indeed to "enlist the cooperation of leading citizens and representatives of every class" in meeting their civic obligation. An anecdote suggests some of the difficulties Stein experienced: after delivering an address to the annual meeting of the New York State Savings and Loan Association, Stein was subjected to a series of charged questions. The strained atmosphere forced the moderator to step into the discussion and assure the gathering that "Mr. Stein has absolutely no interest, whatever, in promoting socialistic policies." When Stein attempted to explain why there was a need for a state-assisted housing program, one board member from a rural district countered by citing empty farmhouses and concluded that immigrants, such as Poles and Italians, simply prefer slum living.[11] Stein came up against laissez-faire ideology—and a little xenophobia for good measure.

The underlying problem, however, was that certain interests that benefited from the system of speculative land development and municipal home rule were threatened by Stein's proposals. The existing system fostered the provincial bourgeosie—including land speculators, small builders, and local businesspeople—who would be cut out from the excessive profits generated by speculative building practice. Consequently, it is not surprising that Stein drew the opposition of the Republican Party, which effectively killed the initiative on the state level. The closest Stein came to success was in 1926, when he secured passage of a bill authorizing a state housing board; but the Republican-controlled legislature failed to pass a bill creating a state housing bank.[12]

In theory Stein's proposal divided the state along well-established political lines, pitting the urban and big capital interests of New York City, represented by the Democratic Party, against the rural, provincial, and upstate capital interests, represented by the Republican Party. Since they were included in the proposal, large, well-financed real

[11]Clarence Stein, "The Savings and Loan Association and Its Relation to the Housing Problem," *Bulletin of the New York State League of Savings and Loan Associations* 3 (November 1925): 3–8.

[12]See Lubove, *Community Planning*, pp. 34–35, 76–77.

estate developers should not have been threatened by Stein's plan; nor were Manhattan real estate interests endangered by a plan that did not take up the question of big-city planning. In fact, one could argue that by drawing off population pressure from New York City, Stein's proposals were perfectly compatible with downtown commercial interests. At least this seems true in theory; in fact, there was no well-organized effort by big capital to push for any plan that would in effect establish the principle of state guidance of the real estate industry. Stein's proposals languished and were forgotten.

Stein and his associates took an approach designed to co-opt business interests. He did not attach RPAA planning ideas to a radical or popular progressive political agenda—an attempt that would have given these crucial planning ideas a necessary popular base of support. Yet given the political terrain of the 1920s any attempt to ground the RPAA in a progressive political agenda would certainly have failed. The larger point is this: one cannot accuse the RPAA, and in particular Clarence Stein, of having failed for want of moderation. Stein faltered not because of the radicalism of the RPAA nor because his politics were those of a naive idealist, but because his policies required changes in the relationship between business and the state that powerful elite segments resisted, in part because of their irresponsibility and lack of vision for the country's future. In effect regional planning of the kind that mattered, the kind that might have reshaped the American metropolis, offended interests that were able to rally support by calling up a laissez-faire ideology. Stein's attempts to work out a Progressive-Era policies for structuring urban development in New York State were attacked in the way World War I housing projects had been: they were seen (incorrectly) as "socialism."[13]

This episode, it is well to remember, is one of the critical turning points in the history of the modern American metropolis. The failure of the RPAA to establish an operating model for housing and regional land use planning in New York State may be contrasted with

[13]One need only consult the *Wall Street Journal* for 18 February 1993 to see that this kind of ideological response is still very much alive. President Bill Clinton's modest economic proposal to shift some spending from the private to the public sphere is greeted by calls of "socialism."

Progressive Era successes in land use planning, most importantly in the assertion of a larger state role in natural resource and scenic lands management. Before these successes, the development of scenic areas, for instance, was not unlike the pattern of urban and suburban development: Niagara Falls, which as early as the 1830s had became a scenic "attraction," became so "cluttered" by "fly-by-night enterprise" that it appeared more like a "cheap circus" than a great scenic landscape.[14] But the steps taken in preserving parks and forestlands were not matched in housing and regional land use planning.

After the political failure in New York State, much of Stein and Wright's attention went into the model housing projects for which the RPAA is well known. The demonstration projects, Sunnyside and Radburn, ensured the RPAA's reputation; they represent the group's most visible legacy. Built to establish the superiority of a "planned community" development, Sunnyside, in Queens, and Radburn, 15 miles from Manhattan in Bergen County, New Jersey, were not complete garden cities. With its suburban location, Radburn was not intended as a model for a satellite city of the future. It did not inlcude such garden city features as a greenbelt, low-income housing, and commercial development incorporated into the residential design. (Commerce was relegated to a corner of Radburn, adjacent to a major railroad line but separated from the residential development by a major highway.)

Both Radburn and Sunnyside, the earlier project, were built by the RPAA-affiliated City Housing Corporation (CHC), a limited-divi-

[14]Dyan Zaslowsky, *These American Lands* (New York: Henry Holt, 1986), p. 13. See also pp. 18–30 and 72–84 on Stephen Mather of the Park Service and Gifford Pinchot of the Forest Service; these men were successful "bureaucratic entrepreneurs." In both cases they enlisted business support: the American Civic Association (a big business lobbying and philanthropic group), the railroads, and the timber companies. This national bourgeoisie was enlisted in the effort not only to overcome provincial opposition to land acquisition but eventually to create powerful federal bureaucracies, the National Park Service and the Forest Service, to retain, acquire, and manage vast federal lands.

dend housing concern established by Stein and real estate magnate Alexander Bing. Despite the CHC's claim in its preliminary study, that its projects would "lead, in all probability, to the formation of similar companies and in a comparatively short time, the country might be dotted with these greatly improved communities," the projects had little effect on changing the way housing was built. Historically, limited-dividend housing companies have had a negligible impact on the total housing stock.[15] Furthermore, the CHC wanted to do more than build housing; it sought to construct garden communities, a much more complicated and expensive undertaking. Little about the CHC housing developments challenged the marginal character of private, semiphilanthropic efforts at housing reform. They have had significance largely as models of town planning.

The first project, Sunnyside, may be more useful today, given our need to redevelop existing urban areas. Since it was built in an area where the infrastructure had already been developed and streets were platted in grids, Stein and Wright had to adapt their planning principles to an existing condition. Humanely scaled buildings were fronted on streets, but in addition they were designed to take advantage of block interiors. Common interior courts and generous playgrounds provided an important off-street environment, a considerable improvement over typical urban apartment buildings (see Figure 10). Sunnyside also had a mix of housing types, from row houses to apartments. This helped to attract a socioeconomically diverse population to the project and thereby realized one goal of the planners.[16]

Radburn did not share the quality of diversity because it was designed as a conventional middle-class suburb, a relatively homoge-

[15]See Daniel Schaffer, *Garden Cities for America: The Radburn Experience* (Philadelphia: Temple University Press, 1982).

[16]Wright later published a book, *Rehousing Urban America* (New York: Columbia University Press, 1935), that bears some looking at because of its practicality on how to reconstruct existing urban residential neighborhoods. Using the planning principles he developed working on Radburn and Sunnyside, Wright advocated reconstructing existing neighborhoods of low-cost housing by razing some structures to provide for badly needed open space.

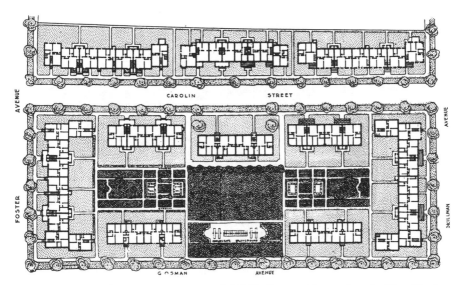

FIGURE 10. The interior courts of CHC's Sunnyside project in New York City were built in relation to the street, but they also provide important off-street space in an urban environment. They have the added advantage of constituting common grounds that give the residents the option to organize the spaces as they see fit.

neous community meant to compete on the suburban housing market. The town's suburban location and lack of industry, its emphasis on detached, single-family houses, and its adoption of the automobile provided a model for a suburban, middle-class town, an idea later adapted by New Town planners in the 1960s.

The most notable features of the Radburn plan were the "superblocks" containing large interior greens and the use of underpasses to ensure the separation of vehicular and pedestrian traffic. Vehicular traffic circulated outside the residential blocks on peripheral roads having access to houses along cul-de-sacs; houses faced inward toward the greens, which were connected by a network of pedestrian paths (see Figure 11). Called the "Radburn idea," the superblocks were intended to make Radburn a "town for the motor age." As Stein explained, the 1920s unleashed "a flood of motors . . . [making] the grid-iron street pattern, which had formed the framework for urban real estate [development] for over a century . . . obsolete. . . . Quiet and peaceful repose disappeared along with safety. Porches faced

FIGURE 11. The second of the CHC projects, Radburn, was typically suburban in the picturesque qualities of its landscape, but it did retain the communitarian concept of the neighborhood center/school and the commons.

bedlams of motor throughways with blocked traffic, honking horns, noxious gases."[17]

One of the purposes in building Radburn was to illustrate how superior residential design could overcome the deterioration of resi-

[17]Clarence Stein, *Toward New Towns for America*, 3rd. ed. (Cambridge: MIT Press, 1971; orig. pub. New York: 1956), p. 41.

dential living in the age of the automobile. It is true that Stein and Wright's solutions, particularly in the design of Radburn, meant a retreat from urban life. But creating sanctuaries from the rush of traffic—places where it is possible to be contemplative, the "quiet squares" of *Communitas*—is essential in creating cities that are fit places in which to live.[18]

The most important innovation of Radburn, however, was the "neighborhood unit," an idea introduced by Clarence Perry, a social worker who often attended RPAA meetings. The idea, which originated in the "community center" movement, concerned the development of the neighborhood as a city unit organized around a central community-based institution, the neighborhood school.[19] In this single respect Radburn was far superior to Sunnyside, which had only its interior courts as a common ground. The common ground of the Radburn community was an institution that had the potential of engaging the residents in their civic life.

Stein had very high hopes for Franklin Roosevelt's New Deal. In a letter written immediately after a private meeting with then governor Roosevelt, Stein said, "He spoke of Russia and how she was calling for large scale imagination in National industrial planning. State planning, yes. Everyone was talking of the planning of the smaller elements. . . . We must [instead] work from the bigger unit down."[20]

There was some hope during the 1930s that garden city housing policy could be realized. Both Stein and Wright became consultants to the Resettlement Administration's Greenbelt Town Program. Under Rexford Tugwell the Resettlement Administration (RA) proposed

[18]Paul and Percival Goodman, *Communitas* (New York: Vintage, 1960; orig. pub. 1948), pp. 164–165.

[19]RPAA, Minutes of Meetings, 1923–1933, meeting of 8–9 October 1927, Stein Papers. Perry was later employed as a consultant for the *Regional Plan of New York*, which was criticized by the RPAA for not going far enough toward regional planning.

[20]Clarence Stein to Aline MacMahon Stein, 24 March 1931, Stein Papers.

that the federal government build hundreds of greenbelt towns throughout the urban regions of the United States. The proposal closely paralleled the RPAA's last policy statement, made before the group disbanded in 1933. "A Housing Policy for the Government" argued that to best meet the need for housing, the federal government should concentrate its efforts on acquiring outlying land instead of expensive sites in the city. The problem with urban sites was that they were overcrowded; the solution was not to tear them down and build higher-density neighborhoods but to create new neighborhoods by building new towns. So the RPAA argued for the building of suburban housing as "community units" and for subsidizing housing so as to avoid "any kind of class segregation."[21] Tugwell hoped that the new towns might provide a more egalitarian and environmental model for suburbanization that at the same time relieved large cities of the pressure of unwanted population. But as Joseph Arnold points out in his study *The New Deal in the Suburbs*, the opposition of provincial elites and the real estate industry, as well as the lack of supportive constituencies, proved decisive. In the 1930s, as in the 1920s in New York State, the New Town idea failed to gain enough political support. In the end political opposition and limited funding resulted in the planning of only six new towns, and of these only three were actually built.[22]

The New Deal was a welter of competing ideas and programs, and the failure of the RA's greenbelt towns opened a broad avenue for conservative New Deal housing policies. These were piecemeal and overly cautious. Although the RPAA, and for that matter many others in the planning profession, understood that most of the future growth would occur outside the urban cores, New Deal housing policy for low income people centered on the cities. Much of the urban housing money was further diluted by conservative interests, which often insisted that funding for housing projects be tied to demolition

[21]"A Housing Policy for the Government, Proposed by the RPAA," RPAA Minutes of Meetings, 1923–1933, Stein Papers.

[22]Joseph Arnold, *The New Deal in the Suburbs* (Columbus: Ohio State University Press, 1971); see also his discussion of the involvement of RPAAers in New Town development for the Resettlement Administration, pp. 191–217.

of slum areas. Politics being the art of the possible, one can certainly argue that given their urban-based electoral strategy, the New Dealers had little choice.[23] But the negative effect was indisputable. By concentrating on urban slum clearance and the construction of large, inner-city housing projects, the New Deal program in public housing became practically irrelevant to shaping the pattern of postwar suburbanization. The urban-oriented New Deal policies neglected the RPAA's alternative development strategy, which lay outside the established urban centers. The New Deal left a legacy of public housing that most Americans were anxious to escape.

In the meantime the private housing industry, which had recently been organized to limit both the economic and geographic scope of federal housing initiatives, became the real architect of postwar (sub)urbanization. While the industry needed government to perfect the financial apparatus necessary to spread homeownership to the working class, it stood against any other form of government intervention. The design elements of garden city planning survived to be used by private developers in the form of the "planned unit development," which popularized curvilinear streets separated by function and the close association of residences with the local school; but the public spaces of the garden city were omitted. The housing industry could and did draw on "traditional" cultural ideals: privacy, the sanctity of the domestic sphere, the isolation of the housewife from society.[24] Consolidated and applied to the context of American

[23]This argument is made by John H. Mollenkopf in *The Contested City* (Princeton: Princeton University Press, 1983). His work emphasizes the power of the urban elements in the New Deal coalition; he also shows that the debate within the New Deal was between an emphasis on construction of public housing and an emphasis on slum clearance. Both were urban-based strategies; the former was favored by the CIO and the progressive middle class, the latter by the AFL, conservative reformers and business interests, particularly the new arm of the real estate industry, the Urban Land Institute. The political difficulty in designing a program for vigorous construction of public housing indicates both how intransigent certain segments of capital remained during the 1930s and how limited was the New Deal.

[24]See Robert Fishman's discussion of the social origins of suburbia in *Bourgeois Utopias* (New York: Basic Books, 1987).

triumphalism following World War II, these cultural ideals became a well-articulated ideology the private housing industry has employed ever since. New Town planning, with its emphasis on multifamily housing units, common spaces, and town centers, was dropped (until a brief and limited revival in the 1960s). So was the possibility of controlling urban sprawl. For whatever elements of "planning" characterized the layout of individual real estate developments (and these varied), the larger hope of the new town, as a way to provide for an orderly and environmentally more acceptable method of handling suburban diffusion disappeared. Very few in the private housing industry cared to contemplate this kind of government intervention. And yet the industry welcomed federal industrial policies in support of suburban housing, policies involving federal loan programs for single-family housing (which were greatly expanded after World War II), and, later, the massive interstate highway system, the biggest boon to sprawling suburban development.

Just as the New Deal's housing and urban development policies failed, so, too, did the major regional planning experiment: the Tennessee Valley Authority (TVA). The TVA echoed Mumford and MacKaye's stress on a specifically *regional* (as opposed to metropolitan) planning; the TVA was concerned with underdeveloped rural regions.

Like MacKaye's emphasis on conserving the "primeval" environment, the TVA plan was intended to encourage soil conservation and reforestation. Furthermore, the TVA would reinvigorate the regional culture by providing for planned industrial development that could bring the benefits of industrialization to one of the most impoverished regions in the United States. It would balance agriculture with decentralized industry, the indigenous regional culture with advanced technical development, and state initiative with cooperation and mutualism. In short, it was the application of planning to the end of coordinating the technical and economic development of a cultural and natural region. As is well known, TVA's foremost responsibility, the key to the whole project, was the building of a series of dams, the first of which, the Norris Dam, included a model new town for housing construction workers. After the completion of the dam, the planned community would remain as the core of what MacKaye called a "regional city." Industries would be attracted by the availability of inexpensive electrical power.

TVA planning should be understood in terms both of Mumford and MacKaye's "regionalism" and of plans for the construction of industrial new towns that had emerged in the 1920s in the United States and the Soviet Union. Industrial new towns were not innovations of the 1920s, and their history has not been socially progressive. Lowell, Massachusetts, and Pullman, Illinois, were 19th-century American examples of social environments controlled by individual companies. Some industrial cities planned in the 1920s also revealed authoritarian social purposes. As an independent government agency, TVA was designed to avoid the negative social effects that usually characterized industry-specific development (the one-industry town).[25] In the specific context of the Tennessee Valley, the TVA was an alternative to Henry Ford's 1922 proposal for a joint government-industry venture along the Tennessee River at Muscle Shoals, Alabama. Adapting a government proposal for the construction of dams suitable for hydroelectric generation, Ford proposed that the power generated be utilized to subsidize a complex of factories producing aluminum, textiles, steel, and auto parts, all of which would be under his control. Interestingly, the plan for Ford's Muscle Shoals site, which called for a seventy five-mile-long linear city with an industrial belt along the river, paralleled the authoritarian planning that held sway in the Soviet Union and led to the creation of the new industrial city of Stalingrad.[26]

[25]On this see, for example, Stanley Buder, *Pullman: An Experiment in Industrial Order and Community Planning* (New York: Oxford University Press, 1967).

[26]In the Soviet Union four industrial new towns, including Stalingrad, were planned and built by the state during the First-Five Year plan (1928–1932). Architect-planner N. A. Milutin conceived his task as building the "ideal socialist city," which turned out to be based on the paradigm of the mass production factory. A city, said Milutin, is like a factory assembly line. Hence Stalingrad was built as a linear city, organized as a series of

By contrast, the early TVA followed Mumford and MacKaye's regionalist concept. The town of Norris, for example, was a planned community built on garden city lines. The TVA also adopted Mumford's argument that healthy regional development should balance industry and agriculture, urban development and rural restoration. Finally, by emphasizing public ownership of the energy infrastructure (power dams) and lands adjacent to the new lakes, TVA hoped to avoid the concentration of power in corporate hands and to heal the Tennessee Valley's environment.

The TVA, a semiautonomous arm of the federal government, had been molded by its idealistic codirectors, Arthur Morgan and David Lilienthal. Sharing the RPAA's hope that advanced technics and economic development could be compatible with regionalism and encourage an ethos of cooperation and mutuality, Morgan hired MacKaye in 1934. But as the TVA grew, it succumbed to the same social relations and economic organization that characterized urban America. In particular, as the larger commercial farmers and industries became more successful economically, they became better organized, more articulate, and better able to bend TVA policy to their liking. Soon the original goals of the TVA, its environmentalism and hopes for autochthonous cultural development, were superseded by a very conventional form of "development." One clear reason for this was the TVA's overcapitalized and centralized technology, which encouraged other large capital investments. TVA be-

---

functional zones, with the industrial zone given prominence. These zones organize the life of the worker, just as they determine the shape of the city. On the Soviet new towns, see Ervin Y. Galantay, *New Towns: Antiquity to Present* (New York: George Braziller, 1975), pp. 42–45. Ford's conceptualization of a linear new town appeared in Littel McClung, "The Seventy-Five Mile City," *Scientific American* 127 (September 1922): 156–157, 213–214. I am indebted to Giorgio Ciucci, "The City in Agrarian Ideology," in Cicuii et al., *The American City: From the Civil War to the New Deal*, trans. B. L. LaPenta (Cambridge: MIT Press, 1979), pp. 336–338, for drawing my attention to Ford's proposal. Ciucci points out that the conception of Muscle Shoals presented by Ford "was directly related to Arturo Soria y Mata's linear city," which emphasized a community of houses, each standing on its own acre of ground.

came the means by which the Tennessee Valley was brought directly into the corporate-dominated national economy. Power dams proliferated, commercial agriculture superseded subsistence farms, migrating industries came to take advantage of cheap labor and federally subsidized power. The original regionalist program Morgan encouraged to support an economy compatible with preserving the land—the establishment of small cooperatives, canneries, creameries, and handicraft outlets—was soon overshadowed by the growth of commercial agriculture. Morgan was forced to resign; a new director of TVA determined to realize a larger commercial value from the enormous investment in infrastructure. Cooperatives and small-scale farmers lost favor. (Significantly, they never had a political power base of their own.) Large-scale commercial farmers courted favor and won it. The New Town program was abandoned; subsequent construction sites had barrackslike structures for workers. The creation of new industries based on agriculture and recreation and cheap power made the valley economically viable, and the hope of developing underdeveloped areas along radically regionalist lines died a quiet death.[27]

[27]Benton MacKaye, "Tennessee—Seed of a National Plan," *Survey Graphic* 22 (May 1933): 251–254, 293–294; reprinted in P. T. Bryant, ed., *From Geography to Geotechnics* (Urbana: University of Illinois Press, 1968), pp. 132–148. The painful uprooting of farm families is described in Michael McDonald and John Muldowny, *TVA and the Dispossessed* (Knoxville: University of Tennessee Press, 1982). Philip Selznick, *TVA and the Grass Roots* (Berkeley: University of California Press, 1949), considers the dynamics of TVA's organization and discusses the conflict between the "agriculturalists" and "super-idealists" in terms of the TVA's mandate to include local people in decisionmaking. The latter group, which included Lilienthal, Morgan, chief forester E. C. M. Richards (and undoubtedly MacKaye, though Selznick does not mention him), sought to reshape the valley in line with the ideals of resource conservation and social progress. The agriculturalists were more successful, however, because their modest goals proved to be easily assimilable by the local elites who were organized around the land grant college system. The immediate issue that divided the two groups was control of the land around the lakes created by the power dams. The idealists considered such land "marginal" for agricultural purposes and wanted to use it for resource and wildlife conservation; the agriculturalists wanted it to remain in private hands. But

In effect Morgan and his associates failed because they were unable to build local and regional communities capable of generating an economy of sufficient power to compete against the state-subsidized and heavily capitalized economic structure. Morgan and Lilienthal's idea of a regionalist economy consisting of marginalized craft and diversified agriculture did not provide a sufficient economic base to compete with the development of agribusiness and the recreational industries. Without an effective political base, Morgan and Lilienthal lost the bureaucratic war for control of the TVA and the political-cultural contest for the allegiance of the Tennessee Valley's people.

Since the TVA, the larger trend has been toward the wholesale abandonment of many regions both industrial and rural. The unregulated "nonmarket" (the power of corporate bureaucracies) has concentrated population and resources on the two coasts. When still in effect, industrial decentralization, now on a very large scale and designed to make use of available natural resources and labor, has tended to be industry-specific and has replicated the social conditions of the one-factory city. This kind of industrial diffusion is still alive: witness the building of new automobile plants in scattered locations away from major urban areas. This strategy is often aimed at undermining existing industrial regions in order to take advantage

---

beyond this question, the two groups fought over the purpose of the TVA and the agency's role in shaping the region's economy. Here the issue of cooperatives becomes fundamental. Morgan and Lilenthal had organized the Tennessee Valley Associated Cooperatives, Inc. (TVAC) as an offshoot of the TVA. TVAC had helped finance several cooperatives and cooperative associations, including the Southern Highlanders, an association of women's cooperatives engaged in handicraft production. This is precisely the kind of culturally appropriate development advocated by MacKaye and Mumford: the wider purpose of cooperatives was to preserve and develop the unique material culture of the valley by finding an appropriate form of economic development. Yet Selznick understood the weaknesses of the TVAC. It was (1) a kind of "random organization" for the ideological purpose of demonstrating the desirability of cooperatives per se; and (2) unrelated to the huge investment in dams and electric generators that was the heart of the TVA's technical program.

of unorganized labor.[28] Industrial diffusion of this kind has no connection to regional resource planning and has become a regressive social and environmental force.

❦

The failure of these progressive New Deal policies reflected the lack of organized constituencies that could exert political power on their behalf. In this sense they were a serious misreading of the way in which bureaucracies operate to augment (and sometimes recontextualize) private interests. Stein should have known better. His attempt at building a coalition of business interests and social reformers around housing and metropolitan decentralization in New York State may have been acquiescent, but it was far more sophisticated than the progressive New Deal programs.

One attempt to address the lack of organized constituencies in favor of progressive planning policies was made by an ex-RPAAer. After her association with the RPAA, Catherine Bauer became involved in the housing movement, eventually accepting the position of secretary of the Housing Labor Conference. During the 1930s, she directed the Carl Mackley Housing Development for the Hosiery Workers Union. The housing was urban based and stood in contrast to the RPAA's touted garden cities. But Bauer had to work within the existing political structures: like New Deal housing projects, the labor-supported housing was oriented toward the inner cities, a weakness that Bauer tolerated because, as she explained to Mumford, it "represented a real gain in understanding, power and responsibility for the people who live in [the urban housing developments]." Believing that people should be more directly involved in making decisions about housing, Bauer was critical of what she saw as the RPAA's inclination to "see 'perfect' housing developments issuing from a supertechnical machine."[29] Bauer concentrated on housing projects

[28]See Barry Bluestone and Bennett Harrison, *The Deindustrialization of America: Plant Closings, Community Abandonment, and the Dismantling of Basic Industry* (New York: Basic Books, 1982).

[29]Catherine Bauer quoted in Donald Miller, *Lewis Mumford: A Life* (New York: Weidenfeld & Nicolson, 1989), p. 334.

conceived by labor unions because they had the popular base the RPAA always lacked.[30] This base worked within the existing structures of power and social organization, which is to say that it had to accept the limitations of federal initiatives. And as we have seen, these were aimed at large urban projects.

To Bauer's credit, in the 1940s she tried to push her constituency to see the importance of providing a suburban alternative to the private housing industry. She led the United Auto Workers in Philadelphia into a coalition with social workers and housing professionals to propose a new pattern of regional development for the postwar era. Arguing that to limit public housing to cities would be to doom it to irrelevancy, Bauer's coalition advocated "regionalism," which it defined as new urban development focused on decentralizing industries and including an important element that would otherwise be missing in suburban growth: public housing designed for the working class. In essence Bauer persuaded the leader of progressive labor unions to support Stein's program for metropolitan decentralization. But the effort was much too late to make political headway, and in overwhelming numbers the membership of the trade unions welcomed the opportunity to purchase suburban houses. The hope for government-planned decentralization gave way to the reality of government-subsidized diffusion on terms entirely beneficial to industrial and real estate interests.[31]

※

This discussion points to one of the significant contingencies of America's development: the goals of federal policies that shaped the

[30]Labor unions that were concerned with broader social questions were just reemerging in the 1930s. They were fragile and had no history of sustained interest in housing issues. Where labor did attempt to speak on housing issues, it concentrated on housing within urbanized areas. See John F. Bauman, "Visions of a Post-War City," in Donald A. Krueckeberg, ed., *Introduction to Planning History in the United States*, (New Brunswick, N.J.: Center for Urban Policy Research of Rutgers University, 1983), pp. 170–189.

[31]See Bauman, "Visions of a Post-War City."

great postwar decentralization of the country's built environment. RPAA environmental planning might have made a considerable difference. Had the United States adopted these strategies in the 1930s, the magnitude of the current environmental crisis would not be as great as it is. RPAA planning might have ameliorated some of the most depressing features of American life: the excessive consumerism, the throwaway culture, the treatment of nature as mere amenity. Had RPAA policies been realized, a different kind of geo-urban structure would have emerged in North America, one that would have offered us considerable environmental benefit. That New York and the nation failed to avail themselves of RPAA ideas is less the fault of the RPAA than the failure of political and economic elites to respond to pressing environmental and human needs.

The RPAA should be appreciated within its historical context. For RPAA predictions were correct: urban diffusion was destined to completely reshape American cities. The pattern was established by the post–World War II suburban boom and ratified by the passage of the Interstate Highway Act in the mid-1950s. "Spread city," or suburban sprawl, was tied to transportation, increasingly to private automobiles, along a vast system of planned, state-subsidized highways. The highways could be justified, as they were by Robert Moses,[32] as a means to escape the city and get to parks and open countryside, but this limited social good was purchased at the unnecessary price of creating a vast (sub)urban sprawl. This kind of planning has permitted the greatest mobility of people and goods ever achieved in human history, but at a staggering environmental and social cost.

RPAA planning, especially as presented by Stein and Wright, pro-

[32]On the influence of the *Regional Plan of New York* on Moses, see Susan S. and Norman Fainstein, "New York City: The Manhattan Business District, 1945–1988," in G. D. Squires, ed., *Unequal Partnerships* (New Brunswick, N.J.: Rutgers University Press, 1989), pp. 59–79, esp. p. 63. The Fainsteins point out that the plan included mass transit, whereas Moses, who had virtually unchecked power as head of a number of state and city agencies, notably the Triborough Bridge and Tunnel Authority, relied exclusively on bridges and highways to transform the city. On Moses see Robert Caro, *The Power Broker: Robert Moses and the Fall of New York* (New York: Knopf, 1974).

posed a different kind of decentralized metropolis: the sprawling character of American cities might have been reined in; the overreliance on the automobile might have been curtailed. If suburbs were going to be built, the Garden City or New Town was the only conceivable alternative to the privatistic suburbs that were constructed after the war. New Town suburbs would have satisfied the people's desire for parklike settings while maintaining some form of public commons and providing for better income-level integration. If suburbs had been designed as garden cities, with integrated functions and well-defined town centers, more of the natural and historical landscape surrounding metropolitan areas could have been preserved and adapted to better use. By establishing something akin to a regional pattern of development, we would have avoided our all-too-frequent habit of obliterating entire landscapes; instead, we would have preserved something of both the biological qualities and cultural patterns of the land itself, while adapting these open spaces to the need of urban dwellers for increased access to recreational areas and local farms. Had development been clustered in garden cities, present-day environmental problems that are related to our overreliance on transportation would be less severe. Current habits of automobile use, generated in large part by suburban strip development along highways, could have been minimized. Instead of organized metropolitan decentralization, we have experienced suburban diffusion. This has made us completely dependent on the automobile and the huge infusions of foreign oil that now sustain the pattern of urban life on this continent.

PART III

Ecological Regionalism: Challenges and Prospects

The erosion of the liberal Center makes it difficult for liberals to undertake even palliative reforms . . . liberal values . . . have come to be embodied in a social order resting on imperialism, elitism, racism, and inhuman acts of technological destruction . . . As a social philosophy, liberalism is dead; and it cannot survive even as a private morality unless it is integrated into a new moral and philosophical synthesis beyond liberalism.

Christopher Lasch
The Agony of the American Left

CHAPTER 9

The RPNY
and the "Ideology of Power"

Like the RPAA, the TVA and the RA attempted regional planning experiments without consolidating either popular or elite bases of political support. At the same time, the greenbelt towns and the early TVA programs presented serious challenges to the established spatial order. This combination of lack of a political base and radical aspirations ensured that the TVA and the greenbelt towns would fail. Because New Deal regional and garden city planning did not develop a discourse capable of defining a "public" interest in regionalist planning, it had no protection against established interests that wanted to limit such planning to an oddity (the greenbelt towns) or, better yet, turn it into an instrument of big capital (the TVA).

As we have seen, in the 1920s regional planning had emerged as a major concern of the profession. And while there were important differences among planners as to the scope of the planning effort, there was general consensus that technological and other structural changes had shifted the locus of planning from the immediate urban area to the wider metropolitan region. Within this consensus important questions remained. To what extent should urban decentralization be managed? How extensively is land use to be regulated? What is to become of the center city? Yet all agreed that planning is essential to the decentralization process. Even the RPAA, which had positioned itself as a proponent of sweeping change, had argued that other regional planners were moving in the right direction.

But by the early 1930s, well before the discouraging events of the New Deal unfolded, Mumford disclosed to his RPAA colleagues that he had major reservations about the planning consensus. His dissension was prompted by the unveiling of the much-anticipated *Regional Plan of New York and Its Environs* (RPNY), published in its completed form in 1931. This plan was a great disappointment to Mumford both in the specific policies it advanced and, perhaps more importantly, in the conception of planning that it promoted. In a two-part article for the *New Republic*, Mumford publicly broke with the *RPNY*.[1] In doing so, he departed from the RPAAs practice, fashioned by Stein and Wright, of establishing a working relationship with mainstream planners. Mumford, in typical RPAA style, did not make public his disagreement with the *RPNY* until he discussed his ideas with the RPAAs inner circle.[2] But Mumford's intention to break with the *RPNY* was very controversial in the RPAA. Charles Ascher, the groups counsel, feared that Mumford's dissension might make the RPAA look "inflexible" and "doctrinaire."[3] It is difficult to know whether Stein or others agreed with Ascher; in any case a year after Mumford's decision to make public his criticism of the *RPNY* the

[1]The "debate" between Adams and Mumford began after the publication of the last volume of the *RPNY*, *The Building of the City* (New York: Regional Plan of New York, 1931). See Mumford, "The Plan of New York," *New Republic* 15 June 1932, pp. 121–126, (22 June 1932): 146–154; and Thomas Adams, "A Communication: In Defense of the Regional Plan," *New Republic*, 6 July 1932, pp. 207–210; reprinted in Carl Sussman, ed., *Planning the Fourth Migration* (Cambridge: MIT Press, 1976), pp. 221–267; henceforth Mumfords article "The Plan of New York" will be referred to as *P of NY*, with Sussman's pagination.

[2]Sussman, *Planning the Fourth Migration*, p. 222.

[3]Letter from Charles Ascher to Mumford, 20 October 1930, Clarence S. Stein Papers, Olin Library, Cornell University, Ithaca, New York.

RPAA ceased to exist. In calling attention to what he saw as the plan's shortcomings, Mumford broke with the supportive approach the RPAA had taken toward the planning professionals and the *RPNY*. Mumford saw the plan as a rejection of both regionalism and decentralization. In his mind this overshadowed the argument that Wright had made in the *CHRP* that the *RPNY* was of strategic value.

In the main the *RPNY* had made a case for a kind of decentralized metropolis, and its vision shared much with what the RPAA had been working to promote among planners, state officials, and builders. The head of the project, Thomas Adams, had been an associate of garden city advocates Raymond Unwin and Ebenezer Howard. With backing from the Russell Sage Foundation, Adams assembled an impressive interdisciplinary team of experts—architects, engineers, economists, social workers, and urban planners—who worked together to produce an integrated plan. But while the *RPNY* appropriated the regionalist *language* of the RPAA, in Mumford's judgment it failed to deliver a regionalist vision. The study represented an advance over typical and limited "highway-and-parks" regional planning. *RPNY* was concerned with the fate of the entire metropolitan region and offered a comprehensive plan for metropolitan decentralization. The plan recognized the need to decentralize population and industry (if not culture and political power), and it called for the construction of garden suburbs (specifically praising Sunnyside and Radburn), the provision of more open spaces, and the coordinated decentralization of industry. It endorsed Clarence Perry's "neighborhood unit," which had guided residential planning at Radburn. Recommending the decentralization of manufacturing and population, the *RPNY* proposed that Manhattan lower its densities while remaining the core of a decentralized metropolitan region united by a multimodal transportation system. Adams envisioned a respatialized metropolis: "our zoning proposals are based on an initial assumption that not more than 40 percent of the gross area of land should be built upon in any part of the city, thus leaving 60 percent open as a precedent to building."[4]

Despite these moves designed to lower densities, Adams assumed

[4]Adams, "In Defense of the Regional Plan," p. 264.

that an inevitable and significant pattern of population growth would force the expansion of the entire metropolitan area. Reading *Building the City*, one finds many historical examples of Adams's axiom that growth is a sign of health. Adams frankly admired capital cities for their concentration of population and power. As the world capital of trade and finance, New York is such a city, and it must remain true to its destiny as a center of power. Such a center will inevitably draw population. But if Adams believed that history teaches us that we must grow or die, he did not present his argument in these terms. He couched the value of New York's future growth in scientific terms: he "proved" it statistically.

The use of science to put what is essentially an historical and valuative position beyond the reach of criticism is an important ploy in Adams's proposal and many subsequent urban plans. While Adams used science to justify his enterprise, in point of fact he had a value-laden vision of greater New York as a urban *region*. This vision was quite distinct from that of the RPAA and certainly very distant from Mumford and MacKaye's sense of a regionalist culture. For while Adams made use of the language of regionalism and proposed to preserve and incorporate natural features into the plan, he was not a regionalist. It is true that Adams put great emphasis on creating integral *places*, but these places—villages, towns, cities—were all to be understood in exclusive reference to the urban center. In effect Adams envisioned the New York region as an ordered metropolis with a planned hierarchy: Manhattan was at the core, surrounded by regional subcenters linked by a multimodal transportation system.

Adams saw the region as an overarching geo-economic structure containing individual units, from large cities to suburbs to towns and villages, each marked by its own "vigorous community life."[5] Yet he considered it essential for localism to be leavened by a "patriotic feeling toward the region as a whole." "Something is lost in local community life," he explained, "when citizens work in one place and have their home life and recreation in another place" (*BC*, pp. 124–125). Nonetheless, this mobility is necessary to a great metropolitan region; economic interdependence is a necessary fact of the life of great cities. But such interdependence need not overcome the

---

[5]Adams, *Building of the City*, p. 122. Henceforth referred to as *BC*.

particularity—"individuality"—of places within the metropolitan whole.

On the surface it appears that Adams made the regionalist perspective an important part of the plan. And there are in fact many fine features of the plan that should have been realized.[6] Adams emphasized open natural spaces and accessible waterfronts; reorienting the city to its rivers would lend a different kind of presence to the city. He even stated that the center of the New York region was not the city but the harbor; it is, after all, this geographic feature to which all the cities in the region owe their existence. Even Mumford praised the *RPNY* for proposals aimed at "improving the waterfront of Manhattan, making it more generally available for recreation at the same time that it is improved for rapid transportation and port uses" (*P of NY*, p. 235). The accessibility of waterfront, the centrality of the port, and the scenic presence of the rivers were all important considerations of the such a plan.

Yet Adams's use of landscape and waterscape was clearly subordinated to an aesthetic vision of the urban core. The natural region provides a *setting* that must be properly employed to build a great city: "the fine features in the environs of the city, the great river . . . the magnificent palisades and wooded hills of the Hudson Valley, give it a natural setting which affords unequalled scope for combining nature and art in a great city" (*BC*, p. 76). The "art" of city building requires many things, but it must begin with attention to natural setting. But a setting is, after all, only a backdrop. Adams's real concerns were the "harmony of style" and "integrity of skyline" of the entire urban core (*BC*, p. 58). He admired the architectural harmony and civic structures of European capital cities and wanted to give New York this aesthetic dimension. Consequently, he endorsed the City Beautiful movement, which featured the "civic center" and other massive monumental buildings standing as "terminal vistas" on prominent avenues.[7] In *Building the City* he maintained that a city's

---

[6] It has taken until now for New York to move to opening the waterfront to the enjoyment of the people, but the new promenade along the Hudson is marred by adjacent redevelopment.

[7] See "The New York Regional Plan," in *International City and Regional Planning Conference* (Baltimore: Norman, Remington, 1925), pp. 212–233, esp. pp. 212, 218.

aesthetic form is a powerful representation of the political and cultural life of a civilization. But a modern (that is 19th- and 20th-century) city such as New York threatens to overwhelm these traditional forms. Harnessing modern technologies of transportation and construction, relatively unrestrained by civic consensus or governmental regulation, the modern city overwhelms the form of a capital or imperial city, a form that evolved over centuries. Adams hoped to reestablish that form and, in the process, to reestablish the ideal of a virtuous elite. His vision was of a gracious and civilized setting for restored "civic virtue."

Adams's vision of New York as "Americas metropolis" was not unlike that of the Progressive Era social theorist Herbert Croly. Croly valued the idea of a European capital as a center of culture:

> Consolidation in social and intellectual matters . . . means the existence of a communicating current of formative ideas and purposes which makes the different parts of the social body articulate, and which stamps the mass of its works with a kindred spirit and direction. The center from which this communicating current of ideas radiates is the social and intellectual metropolis of a country.[8]

The capital city becomes the center of the national life, the figurative head that directs the body. Instead, we have in America "a low organism, full of vitality, but with the vitality resident in its members rather than in its central parts."[9] Adams said we need New York as an intellectual and cultural capital of the United States to create, as in Europe, a nation that is governed by established canons of taste and culture that right the formless and, democratic cultures of America. Adams was working with Crolys vision of a centralized, urban, national culture, and this vision, whatever its merits, was profoundly antiregionalist.

New York was to become a place representative of the whole: the region, the nation, the world—all would look to the city as a representation of great cultural significance, for every other place is pro-

[8]Herbert Croly, "New York as the American Metropolis," *Architectural Record* 13 (March 1903): 193–206, quoted on p. 197.
[9]Ibid., p. 197.

vincial. In short, his was a cosmopolitan not a regionalist vision. Moreover, the social order Adams connected to the visual order of the capital city focused on the national bourgeoisie. Monumental buildings and terminal vistas, he assured readers, arouse the "spirit of the citizens" (*BC*, p. 125) and promote the recovery of the "civic virtue" (*BC*, p. 87) characteristic of the Greek and Roman polis. But who are these "citizens"? They are modern businesspeople. To make the businesspeople true civic-minded citizens, we have to convince them that it is possible to tame the visual disorder they have created without damaging their economic interests. It is not necessary, Adams implied, for the business sector to forsake economic gain for the sake of civic virtue. Adams even accepted the skyscraper, which he knew symbolized the synthesis of modern technology and speculative finance (see Figure 12). Skyscrapers were acceptable to Adams if they were made part of an urban design that provided adequate spatialization, terminal vistas, and other kinds of monumental buildings. The emphasis was on a comprehensive urban design that created "spaciousness in surroundings . . . which is characteristic of great capital cities" (*BC*, p. 75).

The attempt to both accommodate and shape downtown real estate interests constitutes the unacknowledged politics of the *RPNY*. When Adams wrote about the limitations of the U.S. federal system in facilitating planning, he really meant that it was impossible to attack these interests. It is difficult to induce municipalities to apply "adequate public control over land development and sufficient restrictions on bulks of buildings" (*BC*, p. 92). Rather than directly challenging this state of affairs, he hoped to convince business elites that it was in their interest to promote a regional plan.

While fending off critics on the left (like Mumford) by claiming the scientific veracity of the regional plan, Adams appealed to business and political elites. His vision of the metropolitan region both preserved their interests and provided a more aesthetically pleasing and, in the long run, economically viable alternative to what planners now call the "continuation-of-trends" scenario. And from where we stand now, it is difficult not to admit that Adams's rendering of an ordered metropolis seems superior to the distended, formless urban agglomeration we see today.

NEW YORK'S ARTIFICIAL MOUNTAIN RANGE

THE FUTURE CITY OF TOWERS   *E. Maxwell Fry*

FIGURE 12. In *Building the City* Thomas Adams, director of the New York Regional Plan, offered an alternative to the RPAA. Adams's civic vision focused on the monumental city as the nucleus of a metropolitan region.

There are many noteworthy features of Adams's plan—the demarcation of the region as an entity with definite boundaries, the effort to contain what we now call sprawl, and the proposals for shaping a reasonable multimodal transportation system. But the plan came at a very high cost, chiefly the loss of Mumford's ecological emphasis on the region. Adams altogether omitted the idea that the significance of place was in large part the result of its interaction with the natural environment. His vision was culturally and politically conservative; he repudiated the aesthetic and political implications of regionalism. In Adams's plan geographic decentralization ironically focused the cultural life on the urban center.

Mumford believed the New York plan was a betrayal of regionalist principles and a political step backward:

> one must finally judge the Regional Plan not by its separate details but by its *drift*. Thus, the report talks about garden cities but drifts toward further metropolitan centralization; it talks neighborhood planning and better housing but drifts toward our present chaotic methods of supplying both; it talks of objective standards of air and light for building but drifts toward overintensive uses of even suburban areas. (P of NY, p. 227)

Always distrustful of what he considered the overly technical focus of many planners, Mumford saw the RPNY as the product of a "group of technicians" who were unable to make a clearly articulated and unified plan from a hodgepodge of proposals. Furthermore, he charged that for Adams and company, planning would be based largely on the manipulation of statistical evidence and that statistically verified *trends* would determine the planners' course. By resting his case on statistical prediction, Adams modeled planning on the "hard" sciences, with the result that the shape of the city became a natural phenomenon—no longer subject to public debate.

The projection of population has become a standard classroom exercise for today's planning students. But for Mumford the RPNY's exercise in population projection was an arbitrary procedure, a game of statistics "determined by the biological ratio of births over deaths" (*P of NY*, p. 229). While the figures proved to be fairly accurate,[10] Mumford's criticism was directed at the practice of using empirical studies to narrow the depth and direction of the planning study. The manipulation of statistics in this instance only reinforced his sense that the future must be judged by the course of the (relatively recent) past. In contrast to the positivism of the statistical study, Mumford underlined the importance of historical contingency. He asserted that the growth rate of the region would not be "determined by the social and economic forces that have acted in the past" but by cultural and moral and aesthetic choices that we make in the present. Population increases, he pointed out, are also the function of "the very plans and policies that are made to meet an anticipated increase or decrease of population" (*P of NY*, pp. 229–230). The *RPNY*, like any planning document, does not embody an "objective" truth, nor is it "scientific." The plan, in fact, would influence the very future it tried to predict. This is an important critique. In the postwar era the influence of positivism turned planning into a technical, statistical, and *predictive* science and taught planners to picture urban development as a phenomenon inaccessible to public debate.[11]

Even so, Mumford did not completely reject Adams's version of planned metropolitan decentralization. He responded favorably to some of the particulars of land use outlined in the plan. For example, he commented extensively on the planning for parklands, largely approving of the "suggestions for more needed park space" while coming out strongly against draining the Hackensack Meadows: "what the Regional Plan proposes is to swallow for housing and for indus-

---

[10]*RPNY* projected a population of twenty-one million in the New York area by 1965. By 1990 the enormous, thirty-one county New York region (as defined by the Regional Plan Association, Inc., which took up the *RPNY*'s mandate for regional planning) reached a population of 19.4 million. See M. N. Danielson and J. W. Doig, *New York: The Politics of Urban Regional Development* (Berkeley: University of California press, 1982), pp. 3–5.

[11]See Peter Hall, *Urban and Regional Planning* (London: Routledge, 1992).

try the last large potential park and open-water area" (*P of NY*, pp. 236–237). Mumford approved of the sites designated for industrial decentralization but pointed out that these were not matched by "proposals for new decentralized business areas" (*P of NY*, p. 235). He went on to suggest "business subcenters" in upper Manhattan adjacent to the newly opened George Washington Bridge over the Hudson and on Jamaica Bay, Queens. These business centers would represent opportunities for "recentralization" of the existing metropolis around smaller central districts. This is an important idea not only because it suggests a considerable engagement with the problems of the existing metropolis, but because it has become significant today in the rebuilding of cities and suburbs.

Mumford acknowledged, therefore, the potential contribution Adams made to the development of a decentralized metropolis. Yet the *RPNY* remained fundamentally flawed: it "lacks the very essence of a good plan: the sense of alternative possibilities" (*P of NY*, p. 239). First, the plan failed to suggest alternative kinds of decentralization and thus failed to present a regionalist alternative to the metropolis. Consequently, the plan was a premature "compromise" with established business interests. Third, and most critically, the Plan fails as a *text* to create a public. To keep planning from becoming "merely the concern of a profession", Mumford wanted public engagement in the planning process.[12] Accordingly, the planner should begin with housing and the local community and not, as in the regional plan, with the prestige of the metropolis. "Mr. Adams," said Mumford, "dismisses the basic housing problem in this area by simply placing it in the category of a general social problem whose solution lies outside the province of the planner" (*P of NY*, p. 246).

Mumford's critique proposed alternative values and ways of living that the new technologies (and modes of social organization) offered. Paul and Percival Goodman set out such a proposal more explicitly fifteen years later in *Communitas*, favoring new "means of livelihood

[12]Mumford, "Regions—To Live In," *Survey Graphic* 7 (May 1925): 151–152; reprinted in Carl Sussman, ed., *Planning the Fourth Migration* (Cambridge: MIT Press, 1976), pp. 90, 92.

and ways of life" that belong to a new kind of regional city. Mumford's criticism of the *RPNY* was drawn from a similar prospect: regionalism as *both* an ecological community and a democratic republic.

▣

On reading Mumford's critique of the *RPNY*, one is struck by his determined opposition to its main features. He rejected the centralized metropolis in favor of the small regional city, the socially stratified city in favor of organic community, and the megastructure in favor of the small building. Mumford's rhetoric was designed to build an audience for his regionalism and regional planning:

> In geography, the word *region* has a fairly exact connotation; but the Russell Sage planners have used it to describe an arbitrarily chosen area of metropolitan influence. Why was an area with a forty-mile radius from the center taken as the basis of the Survey and the Plan? This question is bound up with all the population predictions and a good part of the plans for transportation and traffic; and yet the intentions and purposes which were in the planners' minds in choosing such an area are nowhere very cogently accounted for. (*P of NY*, p. 228)

There is little attempt to consider the natural regional structures—hydraulic, climactic, or biological—either in defining the area of study or in shaping the regional plan. Instead, the *RPNY* preserved and extended New York's image of power (and conceit) through its acceptance of the unlimited growth of the metropolis, including provisions to permit the erection of still more skyscrapers. For Mumford, the chaos of the "paleotechnic disorder" had no better representative than the skyscraper. In *Sticks and Stones* he portrayed it as a "pure mechanical form" consonant with the unjustified accumulation of power in the large cities.[13] The skyscraper had already become the symbol of New York or, more precisely, a symbolic part of the "myth of [New York's] own greatness." And the myth of New York, as embodied

---

[13]Mumford on skyscrapers in *Sticks and Stones: A Study of American Architecture and Civilization*, 2d ed. (New York: Dover, 1955; orig. pub. (New York: Boni and Liveright, 1924), p. 78.

in the skyscraper, entrenched the metropolitan ideal. Every city rushed to imitate it, and smaller towns sought to imitate the imitators.

His disparagement of the metropolis was as much a critique of a "way of life" (to recall the Goodmans) as it was aesthetic criticism. It echoed his cultural critique of "frontier-mentality" capitalism. Understood in this way, the construction of skyscrapers is the modern equivalent of the clear-cutting of the forest. It is so destructive and chaotic that it consumes urban communities and renders the city unfit for living.

Like Adams, Mumford diagnosed the common symptoms of a city in distress as having their origin in 19th-century paleotechnics. But Mumford was critical of Adams's strategic response to this crisis. Consequently, Mumford repudiated the *cultural premises* of Adams's work, both his appeal to traditional European urban form and his positivistic approach to planning. By contrast, Mumford believed that the concurrence of true "civic virtue" and great inequalities of wealth was unlikely, that a city should be less a monument to the past than a decentralized, living "community." The time had come to decentralize urban form and contextualize culture and politics in a broader regional setting.

Adams praised the capital city of early modern Europe, a subject Mumford turned to in *The Culture of Cities*, published six years after his critique of the *RPNY*. Mumford noted that "consolidation of power in the political capital was accompanied by a loss of power and initiative in local centers: *national prestige* meant the death of local municipal freedom" (*C of C*, p. 80, emphasis added).

As Ferdinand Braudel points out, the early modern urban economy became increasingly dependent on political power and organized violence. Mumford recognized this link between the rise of the new capital cities and the accumulation of economic and political power:

> In the Middle Ages the soldier had been forced to share his power with the craftsman, the merchant, the priest: now, in the politics of absolute states, all law had in effect become martial law. Whoever could finance the army and the arsenal was capable of becoming master of the city. . . . [This] transformation of the art of war gave the baroque rulers a powerful advantage over the real corporations and groups that constitute a community. It did more than any single force to alter the constitution of the city. (*C of C*, pp. 87–88)

Mumford also read this baroque city as a "site of accumulation" that reflected an "ideology of power" (*C of C*, p. 89). Its avenues, the city's "most important symbol," represented not simply the "geometrizing of space," which gives the "appearance of order and power," but the desire for the "conquest of space." More than a symbol, these avenues permitted the military easy access to the heart of the city and created the space for wheeled carriages that entailed the first significant separation of urban classes: "the rich drive: the poor walk" (*C of C*, pp. 94–97).

The capital city came about with a "new conception of space" reduced to "measure and order" and related to capitalism's "abstract love of money and power" (*C of C*, p. 91). Seeking prestige, status, power at the expense of justice, mutuality, and balance, the capital city is not unlike the monumental building. Both represent a "respect for death which is essentially a fear of life" (*C of C*, p. 434). The built environment is always a medium, and the message of the capital city is centralized power. Mumford clearly rejected this vision of the city and of social and political life.

Consequently, Mumford criticized the *RPNYs* ambitious plans for comprehensive new highways and greatly expanded mass transit: "colossal highway and rapid-transit schemes" must be understood as "*an alternative to a community building program*; not a means to it" (*P of NY*, p. 247). His goal was the greater efficiency of living that may be achieved by carefully planning the relation of work and residence. Minimizing daily travel over long distances to get to work, school, or shopping was consistent with the new technics of flexible transportation and electrical power. The logic of such a unified place demands the active intervention of the state. "City development," he wrote, "must be placed, as in Holland, under competent regional and local authorities, who are empowered to purchase land, to design and build and operate new communities" (*C of C*, p. 436).

The power of Mumford's moral and social criticism of the metropolis is evident in his treatment of the capital, or baroque, city in

The Culture of Cities. Here Mumford showed that design is a dependable indicator of the social order. In this sense his treatment of Adams's recovery of the European capital city is in keeping with his cultural critique of Western civilization in *The Golden Day.* Mumford's aesthetic criticism is important, then, because it frames his interpretation of industrial civilization, which he links both to the destruction of nature and the loss of democratic community. His critique of earlier urban forms as well as his vision of a regional democratic community turns on his objection to the concentration of power. This represents a critical shift in Mumford's thinking.

Mumford's critique of the baroque city *begins* as an aesthetic and geographic critique and infers social relations from the spatial relationships created by the structure of the city itself. The abstract spatializations of the capital city, its monumental scale, and its separation of life functions are all related to the increasing social divisions that mark the accumulation of power by new elites. This critique invokes the spirit of *The Golden Day*, the holistic sense of *place* inspired by a culture grounded in the natural region and in a civic spirit. In *The Culture of Cities*, he extended his admiration for the organic design of the New England town to the design features of medieval cities—their "rural character," "usable open spaces," disposition to follow natural contours, streets that function as "footways" connecting little "islands" of buildings, and a pattern of growth that creates "small cities, distributed widely over the landscape" (*C of C*, pp. 43, 44, 56, 59). These places suffer far less from the functional differentiation of space that has characterized the modern city and has led to the separation of living from working. If *The Golden Day* revealed the 19th-century weltanschauung of possessive individualism that justified the opening of the American frontier and the assault against nature, *The Culture of Cities* disclosed the assault on medieval communes ("the real corporations and groups that constitute a community") by the rationalist elites of baroque cities. The newly spatialized cities accompanied economic specialization and social fragmentation. The accumulation of power in joint-stock companies required the destruction of the medieval guild. The mutualism of the early guild made it a true "corporation," a social form Mumford contrasted with the form of "merciless class-competition and individual self-assertion" of capitalism (*C of C*, p. 40).

Mumford extended his organicist method of uniting aesthetics and function to critique the structures of power in highly centralized modern societies. Like Immanuel Wallerstein, Mumford linked social inequities to the hierarchies of geographic development. As Wallerstein has shown, differentiation between "core" and "peripheral" regions constitutes a kind of "economic specialization" that grew out of the city's relation to the countryside; it was, therefore, exploitative by nature and became a necessary component to boosting profits.[14] By the 19th century this "world-system" was well established. Core regions, located in the industrialized world, depended on the exploitation of resources and labor of peripheral regions, both within and outside the industrialized countries.

Consider that virtually at the very moment Mumford contested Adams's imperial vision of New York, New York was emerging as the world's leading city, succeeding London, which had over a century earlier displaced Amsterdam.[15] As Braudel reminds us, great cities depended on the restructuring of the world economy. The "classic sequence" of urban succession began in the late 14th century with the rise of Venice, followed by the successive hegemony of Antwerp, Genoa, Amsterdam, London, and New York. Other cities had their place in this changing hierarchy, which is framed by the differentiated "zones" that surround the leading urban center. The leading city constitutes what Wallerstein calls the "core," and it creates its own "pattern of domination" by utilizing a different combination of "weapons of domination: shipping, trade, industry, credit, and [finally] political power or violence" (*P of W*, p. 35). The resulting world economy "rests upon a dialectic between a market economy developing almost unaided and spontaneously and an

[14]See Immanuel Wallerstein, *The Capitalist World-Economy* (Cambridge: Cambridge University Press, 1979).

[15]Ferdinand Braudel, *The Perspective of the World*, vol. 3 of *Civilization and Capitalism: 15th–18th Century*, trans. Sian Reynolds (New York: Harper & Row, 1984), p. 43; hereafter referred to as *P of W*.

over-arching economy which seizes [the local] humble activities from above, [and] redirects them and holds them at its mercy" (*P of W*, p. 38). This imposition "from above" may be understood as a fundamental structure of power. In Mumford's terms it comes of the capacity to impose an abstract, money-driven (dis)order on the economies of geographic particularity. Braudel shows how this capitalist "nonmarket" exploits the spontaneous market economies and concentrates wealth in an urban hierarchy. "It was the destiny of this local economy . . . to be from time to time absorbed and made part of a 'rational' order in the interest of a dominant city or zone, for perhaps one or two centuries, until another 'organizing centre' emerged; as if the *centralization* and *concentration* of wealth and resources necessarily favoured certain chosen sites of *accumulation*" (*P of W*, p. 36). With "centralization" and the "concentration" of capital, conditions exist for exerting greater and greater influence, for further rationalizing and extending the world economic system oriented around great centers of power. Not only does this lead to the power to reshape production, bringing about various "industrial revolutions" in which imposed organizational and technical capacities transform the way of work, it also helps to generate the accumulation of state power necessary to war.

Mumford posited a decentralized public sphere as a countervailing force to these structures of power. Decentralization entails "more than . . . the [distribution of] overburdened physical plant and equipment of metropolis: [it] means equally the spread and reintegration of the *organs of the common life*" (*P of NY*, p. 256, emphasis added). Mumford would have the reinvention of the commons and the restoration of the mutualist values and practices of a civic life. In *The Culture of Cities*, Mumford recalled New England meeting hall democracy in which citizens "saw and heard their fellow citizens, and . . . discussed problems relating to a *unit* immediately within their grasp and vision" (*C of C*, p. 483, emphasis added). The "unit" that Mumford recovered was the polis.

Mumford extended his aesthetic theory of "place" to a social theory that emphasizes political and social decentralization. "Decentralization," Paul Goodman says, means "increasing the number of centers of decision making and the numbers of initiators of policy; increasing the awareness by individuals of the whole function in which they are involved; and establishing as much face-to-face association with decision makers as possible. People are directly engaged in the function."[16]

In *The Culture of Cities*, Mumford argued that the emphasis on the physical design of neighborhoods, as exemplified by his RPAA colleague Clarence Perry's "neighborhood units," must be broadened. The task, he asserted, was to "organize neighborhoods and corporate organizations" in order to emphasize the "political functions of the community" (*C of C*, p. 483). This is difficult as long as we remain exclusively tied to a conception of parliamentary democracy that is "abstract and disembodied" (*C of C*, p. 483). So Mumford immediately restated the need to "design cities" to make this localism possible, that is, to foster local institutions of self-governance (*C of C*, p. 484). He believed the local sphere challenged the structures of centralized power. Mumford recalled the original goals of the garden city movement: how putting all ownership of land and buildings in the hands of a "common authority" charged with developing the garden city ensures that "such increments [in real estate value] that may arise through the growth of the garden city must be reserved for the community" (*C of C*, p. 396). Furthermore, the local sphere is more than place-centered; it involves

> an active trade-union and co-operative movement: the first to push wages upward, claim a larger share of the total product, and create an effective political demand for government-aided housing: the second to organize and administer the units built, focusing and interpreting the consumers demand, acting as a *mediator* between the official agencies and the professional services and eventual occupants. (*C of C*, p. 471)

Note that not all of the "structures of common life" are place-centered; unions are organized not in response to the politics of place but in response to the inequities implicit in the division of labor and the alienation of consumption and production.

[16]Paul Goodman, "Notes on Decentralization," in Irving Howe, ed., *The Radical Papers* (Garden City, N.Y.: Anchor Books, 1966), p. 190.

Mumford recognized that decentralization could not be absolute. He was aware, for instance, of the tendency to use the neighborhood ideal to justify isolated, smug enclaves of privilege in an unjust society.[17] He knew that as a political practice, decentralization had supported privileged interests and that decentralization may be used to deny that we are united in modern societies by large and remote networks that create functional and economic interdependence, a condition Dewey called the "Great Society."

Knowing this, he acknowledged the need for governance on a large scale: "while regions should become the basic units of political and economic life, the inter-relation of regions within the province, of provinces within the `country,' is no less important: for both cooperations and conflicts must take place over these wider areas" (C of C, p. 363). And in practice he called for all kinds of state intervention in subsidizing housing, controlling land use, and even rerouting trade. Mumford seemed to expect the decentralization of political power to be accompanied by a marked increase in the exercise of state power. This is not in itself contradictory.[18] But Mumford's discussions of the issue remain confusing. Sometimes he wrote about participatory local government displacing representative government (C of C, p. 483); at other times, he writes about supranational "world authority" that was apparently above representative government (C of C, p. 370).

His most consistent position was the one that Dewey took, that representative and participatory forms of democracy should be seen "as supplementary and mutually enhancing."[19] Therefore Mumford

[17]For an excellent contemporary discussion of middle-class "homeowner" movements that have appropriated the language of decentralization, see Mike Davis, *City of Quartz* (London: Verso, 1990), pp. 153–219.

[18]For a contrary view, see Murray Bookchin, *The Rise of Urbanization and the Decline of Citizenship* (San Francisco: Sierra Club Books, 1987). Bookchin insists that the distinction between "the political" and "the statist" is crucial (p. 53). This is to say that the state can never be representative; by its nature it cannot respond to participatory democratic publics.

[19]Robert Westbrook, *John Dewey and American Democracy* (Ithaca: Cornell University Press, 1991), p. 549.

was able to criticize the *RPNY* for failing to engage public participation. As Dewey understood, making the state responsive to the public interest is an ongoing test of democracy.[20] It requires not only the development of democratic publics, but the negotiation of power at the national level. In *The Culture of Cities*, Mumford maintained that any kind of localism or regionalism cannot succeed without the networks of power essential to establishing the public interest. This interest may be expressed in a "service State" necessary to "reapportion the existing balance of power within the `nation,' to equalize the privileges of different regions and groups, and to distribute the benefits of human culture" (*C of C*, p. 364).

[20]Mumford was apparently unaware of Dewey's debate with Walter Lippman and the democratic realists. Lippman defended the state as a creation of elites that operated in the public's interest but without their full understanding. From his point of view, government is best left to elites. Dewey held that Lippman's position reflected the "eclipse of the public" in American life and, furthermore, that the exercise of the public at the local level as a participatory democracy is necessary for any kind of representative government. See ibid., pp. 308–311.

CHAPTER 10

Place and Polity in the "Neotechnic" Era

Mumford's early optimism rested on his hope that "neotechnical" modernization would make possible a new social and cultural order. And if he was unduly optimistic, his concept of "neotechnics"—a technical complex resting on electric power—was hardly a phantasm. As Timothy Luke points out, Mumford's idea of a neotechnical regime "loosely parallels the emergence of Mandel's 'late capitalism,' Galbraith's 'new industrial state,' Baran and Sweezy's 'monopoly capitalism'. . . . Ultimately, it is the basis of the transnationalized, high-technology regime of the global economy after the 1960s."[1] The regime is characterized by certain factors that Mumford also anticipated: the application of science to production and the consequent emergence of the engineer; the geographic diffusion of industry made possible by electric power; the transformation of the worker from machine-tender to machine-supervisor and technician. Luke suggests that Mumford's hope for decentralization derived from his reading

[1]Timothy Luke, *Social Theory and Modernity: Critique, Dissent and Revolution* (Newbury Park, Calif.: Sage Publications, 1990), p. 40.

187

of neotechnics as a revolutionary method of production: "it was [now] feasible to decentralize the labor process to approximate a more 'humane' mode of living. City and country . . . tended to reintegrate and become more contemporary."[2] As Luke observes, Mumford looked at neotechnics in terms of the possibilities it presented for political and social change.

Thus in *The City in History*, Mumford again evoked the polis of ancient Greece, the "dream of Jeffersonian democracy," and the sense of "neighborly life" that would make possible the restoration of the "small face-to-face community of identifiable people, participating in the common life as equals."[3] And he spoke for democracy; local and regional efforts in planning should build the structures of civic participation necessary to real democratic decisionmaking. But Mumford never addressed the difficult question of how to revive the "civic sphere" and "sense of place" in the face of new and increasingly intricate networks of economic and political power created by neotechnics.

He did realize that it was foolish to think that large governmental structures could be supplanted by local and regional ones. Like Dewey, he saw that the modern world generates levels of interdependency that require management by centralized institutions. Thus the conundrum: local and regional initiatives are more flexible and potentially open to democratic process, but centralized power is necessary to functional interdependence. Mumford left us with important, but not entirely articulated, responses: (1) only by reviving local and regional public spheres and generating innovative approaches to our problems is it possible to make national and international structures responsive to the will of the people; but also (2) the revival of the local and regional public spheres cannot be independent of—in fact must be in relation to—the centralized networks of power. In short, to put it in terms of American politics, local and regional planning cannot come from the federal government, but neither can any local initiative ignore the power of the federal government to shape the basic conditions under which we operate.

[2]Ibid., p. 42.
[3]Mumford, *The City in History* (New York: Harcourt, Brace, 1961), p. 500; hereafter referred to as *C in H*.

Increasingly, then, Mumford understood the need for challenging and reshaping existing institutional bases of power. But even though he could no longer believe that perception—operating through our individual and collective imagination—was sufficient to the transformation he had once prophesied, he rightly refused to think that a change in perception had become irrelevant. The organicist prospect of *ecological regionalism* was still important. The master key to understanding Mumford as an intellectual is to correctly view his organicism. His interpretation of technological innovation is best understood in these terms: neotechnics is a metaphor for a new way of living. It provided an opportunity to reenact our collective life and thereby re-create the relation of culture to nature. The possibilities inherent in a decentralized neotechnics could provide "occasions of experience" that allow us to imagine and create a new relation between production and consumption, between function and aesthetics. Decentralization has the potential of overcoming the ill effects of the fragmentation of space into functional spheres, a consequence of the centralization of production under paleotechnics. These ideas reflect Mumford's effort to find the interrelation of all things and to uncover what Whitehead calls the "fundamental unities" of existence in "occasions of experience."

The real significance of Mumford's work lies, therefore, in his capacity to reimagine "production" in relation to place, both civic and ecological, and thereby remind us that it is possible to create "a finer relation" to our environment. He had little to say about how these necessary acts of imagination could actually challenge the structures of power that underlie our present relation to nature, though, as we have seen, he became increasingly aware of the need to do so. This is to say that while Mumford never worked out a social theory, he continued to search for the political implications of his ideas. This was not the case before the break with Thomas Adams and the *RPNY*. In his early work he thought that the organicist possibilities of neotechnics might evolve—outside of politics—through the evolution of science itself. By the early 1930s his debate with Adams demonstrated the obvious limitations of science and the potential for the

misappropriation of scientific certainty. As a way to package expertise, science has become the tool of managers and professionals, a
means of consolidating their exclusive hold on power by limiting
public debate. This is why Mumford came to insist that the most
important characteristic of a planning document should be its *contingency*. Let the affected communities see multiple possibilities at
hand. The planning professional presents different options, and planning becomes a way to address—even a means to create—the public. Open planning to public investigation: make it subject to the
practice of what MacKaye calls "exploration." Such thinking is
organicist because it trusts the possibilities inherent in experience. It is
democratic because it will reinvigorate public life. Planning should build
on a shared sense of responsibility. Urban and regional design cannot be
separated from the capacity to create a public sphere, a commons, and
the sense of place associated with commonality and community.

The "facts" of postwar urban development forced Mumford to con-
front the depth of our political and ecological dilemma. The promise
of garden city regionalism gave way to the reality of the "anti-city."
The "anti-city" was Mumford's description of the new suburbia, a
neotechnical geography that failed to find appropriate forms or structures. By the late 1950s, when Mumford substantially revised *The
Culture of Cities* as *The City in History*, the diffusion of population,
commerce, and industry had transformed the prewar American metropolis. The development of a vast system of federally subsidized
highways encouraged the rise of the trucking industry and nearly
universal automobile ownership. This newfound "hyper-mobility"
transformed the centralized railroad and streetcar metropolis of the
turn of the century into a *fragmented metropolis*, or, as its defenders
called it, "megalopolis." While the fragmented metropolis may be
understood as the culmination of a process of differentiating urban
functions that began in the premodern era, in form and function the
megalopolis differs from anything that preceded it. It reflects the
decentralizing tendency of neotechnics, creating a new geography.
The centralized metropolis of the early 20th century was now sub-

merged. The centralized metropolis had a well-defined hierarchy of places. Business activity and commerce had been heavily concentrated in the downtown, industries anchored in "factory districts," and residences concentrated by class in urban neighborhoods and railroad suburbs. The new pattern of movement made possible by the neotechnical revolution ignored such hierarchies of place. The resulting diffused city lacks a coherent center and is linked by a system of highways that bypass the established core areas.[4] Workplaces— everything from manufacturing plants to office buildings—are increasingly located adjacent to highways. This fragmented metropolis has, as Robert Fishman points out, made the "traditional suburb"— a bedroom community that remained "completely dependent on the central city for jobs and essential services"—an antiquity. Thus, what the RPAA in its day understood as a metropolis has been superseded by a new urban form, a "vast super city" consisting of dispersed urban functions, united by a network of superhighways and predicated on automobile transportation. This diffused city is so divergent from older suburban forms that Fishman calls it "technoburbia."[5]

[4]As Scott Lash and John Urry, *The End of Organized Capitalism* (Madison: University of Wisconsin Press, 1987); and E. Schoenberger, "From Fordism to Flexible Accumulation," *Environment and Planning* 6 (September 1988): 321–338, argue, the decline of established urban hierarchies parallels the rise of post-Fordist production. They maintain that in the past three decades a new kind of geo-economy has taken shape. The era of "organized capitalism" or "Fordism," which emerged by the 1920s by linking production-line industrialization to mass consumption, has been displaced by a new regime driven by telecommunications and computer-assisted design. The results are the reduction of size and scale of manufacturing, greater functional differentiation in manufacturing, geographic dispersal of the manufacturing process. Thus geographically we see the rise of forces that tend to disperse and decentralize production; the movement is from "high-volume standardized production" to "flexible system production." Mass production in large-scale plants and bureaucratic centralization—features of "Fordist" production—are being rethought.

[5]Robert Fishman, "America's New City: Megalopolis Unbound," *Wilson Quarterly* (Winter 1990): 25–45, quote on 27. Fishman also calls this a "decentralized" city; I prefer "diffused" city, to distinguish our present sprawl from the ordered, garden city decentralization advocated by Mumford and the RPAA.

Mumford found the unfolding of "technoburbia" sobering and frightful. In *The City in History*, he called it an "anti-city." It is created by forces of dispersal that seem to operate outside of human control and have

> automatically pumped highways and motor cars and real estate developments into the open country [and accordingly] have produced the formless urban exudation. Those who are using verbal magic to turn this conglomeration into an organic entity are only fooling themselves. To call the resulting mass Megalopolis, or to suggest that the change in spatial scale, with swift transportation, in itself is sufficient to produce a new and better urban form, is to overlook the complex nature of the city. The actual coalescence of urban tissue that is now taken by many sociologists to be a final stage in city development, is not in fact a new sort of city, but an anti-city. As in the concept of anti-matter, the anti-city annihilates the city whenever it collides with it. (*C in H*, p. 505)

Mumford's comment restates the RPAA's position that without careful planning the new technologies of decentralization would create a formless exurban sprawl.[6] The RPAA had always insisted that highway development be tied to land use planning; the group had promoted the garden city as a new town capable of absorbing population growth outside of urban centers while preserving the value and character of the surrounding countryside.'

Still, it is apparent that Mumford had recontextualized the issue of technology and its potential. In the 1920s he had characterized the new technologies that permitted diffusion (the automobile, the radio, the electric power grid) as a "revolution," the value of which was confirmed by the "fourth migration," the "tidal movement of population" out of the central cities (see Chapter 5). The kind of planning the RPAA proposed was presented as a natural accompaniment to an ongoing process of technological innovation and social evolution (adjustment). Mumford believed that this process employed the potential for human development. As an alternative to existing arrange-

[6]In part Mumford was responding to the argument put forth by Jean Gottman in *Megalopolis: The Urbanized Northeastern Seaboard of the United States* (Cambridge: MIT Press, 1961) that the new metropolis was simply a new urban form.

ments, it reflected the need to correct the overcentralization of population and activity in the urban centers and followed from human ingenuity and intelligence. His early appraisal was optimistic, as if the invention of the technologies and the increasing suburbanization of the population were prima facie evidence of the suitability of the RPAA vision.

By the time he wrote *The City in History*, Mumford had resituated the "fourth migration." He stopped thinking of it in terms of *what might be gained* and approached suburbanization in terms of *what has been lost*. This indicates the extent to which Mumford and the RPAA had relied on a theory of modernization, a process they heralded as opening a new age. But the new age had gone awry. Even the application of many planning practices associated with the garden city were trivialized by the creation of planned suburban "new towns." As the number of these new towns grew—in the early 1960s several were begun under private auspices[7]—Mumford became increasingly disenchanted with them. "They are merely suburbs dressed up to look like cities," he complained to British garden city planner Frederic Osborn. As "distinctly [and] purely middle class communities," they excluded "workers in the lower ranks." Furthermore, the exclusive emphasis on design resulted in "swank residential establishments," no substitute for creating genuine urban centers. Rather than founding communities, New Town planners had found a way to represent the values of metropolis: privacy, consumption, and escape. New towns lacked urbanity; they emphasized the garden to the exclusion of the city: "visually the garden displaced the city." This reduced densities, and as a result "compactness," a feature necessary to encourage "daily encounters and mixtures among people," was lost.[8] The new towns, moreover, became less and less distinguishable from the rest of the "anti-city."

"I cannot share your optimism about our present state," Mumford

---

[7]Notably two in the suburban environs of Washington, D.C.: Reston, Virginia and Columbia, Maryland. In 1965, Congress authorized the New Town Act, a mere gesture at the social and environmental goals set forth by the RPAA. See Mel Scott, *American City Planning since 1890* (Berkeley: University of California Press, 1969), pp. 644–648.

[8]Mumford, *The Urban Prospect* (New York: Harcourt, Brace, 1968), pp. 150–152.

admitted to Osborn, "but I have always granted the scientific possibility of miracles."[9] One cannot help but be impressed by Mumford's
hopefulness in the face of the perception that technological innovation and planning, upon which he had placed so much importance,
had apparently been subsumed in the "megamachine." The tempta-
tion to conclude that it was all beside the point was always present.
In the final chapter of *The City in History*, Mumford quoted Henry
Adams, who speculated that the perfection of the machine would
wear down, fragment, and finally collapse civilization: "it is useless
to speculate about the future of cities until we have reckoned with
the forces of annihilation . . . [that] are working to bring a more
general breakdown" (*C in H*, p. 558). But he went on to say that "if
the picture were as grim as that I have painted in this chapter, there
would be no excuse for writing this book" (*C in H*, p. 560).

What Mumford experienced in the act of writing yet another book
on the evolution of cities was akin to a crisis of faith. He could no
longer believe, as he had in *The Culture of Cities*, that the "prophetic"
emergence of new technics in architecture, in community and regional planning, and in the means of production itself was moving
us toward a "biotechnic society: a society whose productive system
and consumptive demands will be directed toward the maximum
possible nurture" (*C of C*, p. 415) of the individual, of the group, of
nature. Instead, it had become clear to him that the neotechnical
revolution had produced dramatically different results. The radical
principle of (geographic) decentralization inherent in the production process of neotechnics had become the basis for the spread of
the anti-city. This realization dismayed Mumford not only because it
meant that the destruction of nature and the natural region was accelerating but because it called into question the framing of
neotechnics in his thought. He had simply assumed that neotechnics,
representative of both a technological complex of technologies and a
new class of "engineers," would create a "bio-technic" civilization.

Mumford's doubts grew in response to the failure of America, in
the crucial decades of the 1920s and 1930s, to address the funda-

[9]Mumford to Osborne, 29 September 1969, in Michael R. Hughes, ed.,
The Letters of Lewis Mumford and Frederic J. Osborn (Bath: Adams and Dart,
1971), pp. 461–462.

mental patterns and practices of urban development (and beneath this the pattern and pace of technological change). These failures— Stein's effort in New York State, the TVA, and the greenbelt town program—became, as Whitehead puts it, "occasions of experience" that helped Mumford shape his continuing search for "creative advance." Thus, when the unfolding of this historical process had discredited his assumptions about the relation of neotechnics to ecological regionalism, Mumford sought other ways to posit the necessary "unity" of experience.

As it became clearer that the development of neotechnics had not produced a regional garden city civilization but had instead set free the forces that created the anti-city, Mumford was forced to rethink his earlier assessment of technological change. The reframing of the history of U.S. urban development—from a narrative of hope in *The Culture of Cities* to one of loss and lament in *The City in History*—was part of this process of reassessment. And he continued this work in the two-volume work *The Myth of the Machine*. His conclusions about the relationship between technology and culture might be compared to those William Rueckert attributes to literary critic Kenneth Burke:

> Human beings' specific genius is for symbol-using; the final genius of symbol-using is no longer the god-head, but high technology is; in and by means of high technology, humans have arrived at the ultimate secrets of life and matter and so have the capability of creating and manipulating new life and of destroying life itself, or at least of all human life, and perhaps the planet earth as well. Humans . . . have become the god-head because they have, as a result of their genius—their desire to know and understand and control everything—arrived at the knowledge that gives them the powers once attributed to God.[10]

Like Burke, Mumford came to emphasize the concentration of *power* that technological innovation makes possible, and, like Burke, he came to realize that "more technology will not solve the problems

[10]William H. Rueckert, *Encounters with Kenneth Burke* (Champaign-Urbana: University of Illinois Press, 1994), p. 85.

already created by technology" and that a "neohumanistic corrective vision" has the potential to overcome our worship of the machine.[11]

Mumford understood, therefore, the importance of "vision" in the Burkean sense, as a moral force, a persuasive implement wielded by the keepers of the nation's social conscience. In this sense "vision" is corollary to "symbol-using" and symbol-making—a necessary and innate part of culture, a process that in modernity, as Rueckert tells us, is displaced from the sacred to the profane. Mumford's "vision" was conservative in that he sought to *recover* what Gary Snyder calls "the poetics of the earth."[12] Mumford retrieved the symbols of regionalism and nature to creatively redefine a sense of "place." His larger lifework—that which framed his interest in cities and planning—was to uncover the sources and the traditions necessary to these fundamental restructurings. But Mumford also came to see that the vision of ecological regionalism had an important political implication. As much as he would agree with Burke that the failure to understand the implications of technological power is a failing of conscience—a displacement of our "symbol-using" power—Mumford also insisted that it is a defect in the very structures of power upon which we've built our polity. Succinctly said: the concentration of power in late modernity corresponds to the increasing trivialization of the civic sphere. Planning must become a way to help redress that imbalance of power and thereby recover the civic sphere. As he came to realize this, Mumford saw that the value of recovering the Emersonian tradition in American letters—with its vision of culture in relation to nature—depended on finding/reinventing a civic or public sphere. At the same time, as Mumford understood it, the polis (the poli-tical), must be broadened to incorporate the necessary concern with aesthetics and production, with what Percival Goodman calls the "double e"—economy and ecology—as productive of an aesthetic-moral *practice* of place.[13]

[11]Ibid., p. 86.

[12]Gary Snyder, *The Old Ways* (San Francisco: City Light Books, 1977), p. 40.

[13]Percival Goodman, *The Double E* (Garden City, N.Y.: Anchor Press, 1977).

Mumford's lifework of joining an aesthetic vision of place and region with a civic social vision came under attack in the 1960s. His sense of the public sphere and its relation to spatial organization was affirmed and at the same time severely challenged in 1961 by the publication of Jane Jacobs's *Death and Life of Great American Cities*.[14] This book heralded an important shift among American intellectuals in their attitude toward the city. The hope of Jacobs's book is, as William Whyte puts it, that the urban "center" might be "rediscovered"[15] and that in the process the movement outward to suburbia could somehow be reversed.

In terms of urban planning theory, Jacobs's work presaged a new paradigm generally called "advocacy planning." Physical planning is seen as less important than "planning for people"; the neighborhood, the starting point of all planning, is less a physical structure than a social one. The planner becomes an advocate for the dispossessed and lines up with social movements.

Consequently, Jacobs was less concerned with the city as a physical structure than with "city life." The essence of city life is found in the neighborhood. But Jacobs's idea of neighborhoods was not well defined geographically; she called them "street neighborhoods" and said that they "have no beginnings and ends setting them apart as distinct units" (*D & L*, p. 120). Instead, she argued that we should understand that real "street neighborhoods," as opposed to the formal neighborhoods conceived by planners, are ad hoc "networks" that overlap and interweave. While they are geophysical in nature, "street neighborhoods" are not "discrete units" (*D & L*, p. 121); they have informal and fluid boundaries that depend primarily on human interrelations, not spatial dynamics. Jacobs's concern was more with what constitutes neighborliness within the social relations of urban life than with formal conceptions of neighborhood. She thus

[14]Jane Jacobs, *The Death and Life of Great American Cities* (New York: Vintage 1963; orig. pub. 1961); subsequently referred to as *D & L*.

[15]William Whyte, *City: Rediscovering the Center* (New York: Doubleday, 1988).

argued that some of the most important forms of neighborliness in the city have no connection whatsoever to geography. These are the various voluntary organizations that draw people from all parts of the city to pursue common interests.

Jacobs defined the street neighborhood as well as the city and the communities of interest by their social functions. And when she wrote of decentralizing political power and reviving the civic life of cities, she thought in functional terms: the "city district" must be large enough (100,000 in a big city such as New York) to exercise political power and to create a framework capable of encompassing the multitude of churches, civic leagues, business associations, political clubs, and other organizations that constitute the real social and political life of cities. For Jacobs, decentralization is a matter of political function, not a question of aesthetic form.

Her book was one of a number of studies of the city, including Mumford's *City in History* and Whyte's *Exploding Metropolis*, that took as their theme the crisis of urban life.[16] But for our purposes here the most important analogue to Jacobs's work was Lucia and Morton White's study, *The Intellectual Versus the City*. The Whites advanced the thesis that the miserable fate of American cities was to some extent due to a deep-seated and pervasive anti-urban bias in American intellectual life. The resulting bias, they said, explained many of the nation's social problems. This thesis was widely accepted in some circles.[17]

Jacobs reminded us that cities could be seen as valued places in their own right, places we have largely taken for granted and by doing so have endangered. Many planners and critics of urban life have focused on the movement out of the city and, accordingly, have neglected the city itself. They have failed to see the valuable social life provided by industrial cities, and, in doing so, they have encouraged the loss of urbanism and the destruction of the city. This argument is a powerful one, a reasonable response to the spread of the anti-city and the crisis of urban life. As a result, we now commonly see cities as endangered by suburbanization and diffusion.

[16]See Russell Jacoby, *The Last Intellectuals: American Culture in the Age of Academe* (New York: Farrar, Straus and Giroux, 1987), pp. 57–61.

[17]Morton and Lucia White, *The Intellectual Versus the City* (New York: Norton, 1962); see discussion in Chapter 6.

Jacobs expressed the experience of a generation that looked at modernity and saw horror: the horror of a technocratic war (Vietnam), technology used by "reasonable men"—*planners of war*—as an instrument against people. She captured this sentiment by representing the destruction of urbanism as another example of planned destruction. Because she is an activist, her account of the evolution of the city had much in common with New Left analyses of the modern liberal state. Both emphasize the role of political elites in defining the structures of modernity, structures the New Leftists consider oppressive.[18] For Jacobs, the elite in question is the urban planners, technocrats who operate under the cloak of scientific "objectivity" but in effect carry out a program of destruction. Planners, she believed, led the assault on the old city. They push "urban renewal" and inner-city highway construction, oblivious, it seems, to the value of existing urban neighborhoods and to the intricacies of urban life.[19]

Jacobs contended that the explanation of the ongoing process of wholesale urban destruction lies with the underlying rationale of classical urban planning. The fault is an almost exclusive emphasis on physical planning, which reflects the planners' false assumptions about urban social life. Planners simply fail to understand cities as social entities because they are schooled in a (faulty) sociological theory of the mass society. Consequently, they propose a spatial order to compensate for the supposed inadequacies of urban life. But these planners, Jacobs argued, fail to understand that an authentic community life exists in the city.

Jacobs proposed that planners begin by reexamining urban society. She rejected the sympathetic view of agrarian communities characteristic of many intellectuals, including Mumford. She believed that the idea that something has been lost to urbanization obscures the potential of existing urban social life to sustain its own forms of community. This important argument is grounded in the approach to

[18]Two important and early examples of such New Left history are James Weinstein, *The Corporate Ideal in the Liberal State: 1900–1918* (Boston: Beacon Press, 1968); and Gabriel Kolko, *The Triumph of Conservatism* (New York: Free Press, 1963).

[19]For an example of this kind of planning, see Robert Caro, *The Power Broker: Robert Moses and the Fall of New York* (New York: Vintage, 1974).

society and social change characteristic of an entire generation of activist-reformers. Jacobs's understanding of the city is based on the experience of a generation that saw the key to political change in social movements such as the civil rights movement and the protests against the Vietnam war.

The connection between the idea of the issue-oriented social movements and Jacobs's urban theory is evident: the New Leftist idea of organizing from below around issues of obvious concern to social constituencies depended on identifying and mobilizing those constituencies. What Jacobs objected to in the reigning social theories of the city was the denial of the existence of constituencies capable of exerting pressure from below. The strength of Jacobs's assertions about urban life depend on (and derive from) the possibility of organizing people to express their interests. She offered an urban sociology and a direction for planning congruent with an organizing strategy based on the existing social world of the "street neighborhood."

There is much to admire in Jacobs's stance, for it follows from her experience as a community organizer and in this sense is pragmatic. Her work embodies a generational experience that led to the rewriting of American social history. In calling attention to the social life of cities, Jacobs reminded us that if "urbanism" contains the conditions that destroy a common life, the urban situation retains the potential to create new forms of social life. Although urban life creates "anomie" (a social void), the urban world can also nurture new kinds of social interaction. In arguing this point, Jacobs avoided the tendency of some observers of cities (from whose ranks we can exclude Mumford) who by employing the gemeinschaft-gesellschaft distinction became mired in a state of moral ambiguity: longing for a past that could not be transferred to the present, seeing "community" and "society" as irreconcilable poles, creating a dilemma that pits intimacy against freedom, community against autonomy.[20]

This is the position of the "democratic realists," liberals such as

[20]See Christopher Lasch's criticism of classical sociology in *The True and Only Heaven* (New York: Norton, 1991), esp. pp. 146–147.

Walter Lippman who defended the rise of professionalism and elitism as the only viable (functional) responses to modernization.[21] Only powerful institutions are capable of overcoming the divisive effects of industrialization and modernization. Thus reform comes through the state. Planning, in this sense, is imposed by professionals whose expert knowledge cannot be shared with citizens, and it becomes the only effective means of reform. Jacobs was right to object to this position.

We can see how far Jacobs traveled from this conception of the urban condition by looking at the following passage from Louis Wirth's influential 1938 article, "Urbanism as a Way of Life"[22]:

> Being reduced to a stage of virtual impotence as an individual, the urbanite is bound to exert himself by joining with others of similar interest into organized groups to obtain his ends. This results in the enormous multiplication of voluntary organizations directed toward as great a variety of objectives as there are human needs and interests. While on the one hand the traditional ties of human association are weakened, urban existence involves a much greater degree of interdependence between man and man and a more complicated, fragile, and volatile form of mutual interrelations.[23]

These "fragile" social relations obviously lack the solidarity of a traditional community. Wirth demonstrated that urban social relations

[21]See Robert Westbrook, *John Dewey and American Democracy* (Ithaca: Cornell University Press, 1991), pp. 294–300.

[22]On Wirth, see Thomas Bender, *Community and Social Change in America* (New Brunswick, N.J.: Rutgers University Press, 1978).

[23]Wirth's "Urbanism as a Way of Life," *American Journal of Sociology* 44 (July 1938): 1–24, quotation on p. 22, emphasis added, became the classic work of American sociology on the city. It overturned the Chicago school, which had dominated the sociology of the city in the United States. Developed in the 1920s, the theories of Robert Park and Ernest W. Burgess treated the city as a spatial phenomenon. The structure of the city, an association of various "communities," is the outgrowth of a spatial dynamic organized around the "mechanism" of "competition": the results are "natural communities" defined by land values. See Robert E. Park, "Human Ecology," *American Journal of Sociology* 42 (July 1936): 1–15. Thus in effect Wirth shifted the discourse away from the spatial organization of the city and toward consideration of its social networks.

need not be atomistic and at times can be characterized by mutuality—"the enormous multiplication of voluntary organizations"—even though not the mutuality of the community of the traditional village. Jacobs found this kind of urban voluntarism and mutualism in New York neighborhoods and saw in them the basis for a new kind of a decentralist politics and planning.

Armed with these sociological insights, Jacobs examined the urban planning tradition. In a critical move that continues to influence many accounts of planning and cities,[24] Jacobs asserted that by concentrating on spatial organization, planners have willfully misread the actual social life of cities. Rather than distinguishing among conceptions of planning, Jacobs saw an undifferentiated planning tradition rooted in the utopian vision of 19th-century architects and planners:

> Nineteenth-century Utopians, with their rejection of urbanized society, and with their inheritance of eighteenth-century romanticism about the nobility and simplicity of "natural" or primitive man, were much attracted to the idea of simple environments that were works of art by harmonious consensus. To get back to this condition has been one of the hopes incorporated in our [planning] tradition of Utopian reform. (*D & L*, p. 374)

All planning, everything from the city beautiful to the garden city and Le Corbusier's plans for a *ville nouvelle*, "always were primarily architectural design cults" (*D & L*, p. 375). They have nothing to do with real urban "communities" or neighborhoods, and as a consequence planning became a weapon aimed against cities.

Rather than examining the theoretical positions from which these conceptions of planning came, Jacobs equated Le Corbusier and Howard, Thomas Adams and Mumford. She did not consider the

---

[24]See, for example, Elizabeth Wilson, *The Sphinx in the City: Urban Life, the Control of Disorder, and Women* (Berkeley: University of California Press, 1991).

actual positions of the people she described, failing, for example, to note Clarence Stein's interest in urban neighborhoods and Mumford's endorsement of community ownership in *The Culture of Cities*. Jacobs chose to ignore Mumford's position on the need to develop a "public"—as "community"—to provide a democratic context for planning practice; she was also unfamiliar with John Dewey's democratic progressivism, his hope (which was substantively Mumford's as well) that there would be room for the small community and local, face-to-face democracy, even within an urban society that because of scale and organization required state planning.[25] Such theoretical matters are of little concern to Jacobs because she insisted on the priority of a politics of organizing. She considered politics of this kind fundamental and sufficient; everything else is distraction. Among the most seductive distractions is the planners' concern with the past. All forms of retrospective vision, she said, are inherently reactionary. All lead to an attempt to privilege art over life.

Ironically, as Marshall Berman pointed out in *All That's Solid Melts into Air*, Jacobs herself engaged in her own version of a retrospective vision. Her conception of the urban neighborhood was rooted, Berman told us, in the early modern city. Her sense of the vitality of the city as a place and her reading of the possibility presented by urban life—the idea of neighborhood diversity within a larger shared social field—were both cultural possibilities offered by the modern

[25]In terms of classical sociological theory, as advanced by Ferdinand Tönnies and Georg Simmel, the Dewey-Mumford position took seriously their argument about the decline of "gemeinschaft" (community) and the rise of "gesellschaft" (society) as an historical development. But Dewey (and Mumford) did not see this line of historical development as inevitable or ironclad; the decline of "traditional" community presents opportunities for the creation of a new kind of community. We can differentiate this position from *both* that of the "democratic realists," who see historical development as a foreclosure of forms of community, and more recent social theorists who deny any relationship between community and historical development. They see, instead, *gemeinschaft* and *gesellschaft* as social tendencies that wane and wax in significance across time. In arguing this position, Bender, *Community and Social Change*, presents a critical review of the entire literature.

city. And Berman argued that many New York intellectuals of his generation, who shared Jacobs's estimation of the experiential value of the urban neighborhood, had yet to come to terms with their personal relationship to the city, to admit that in many cases they left those neighborhoods to pursue their own careers.[26] This is to say that the neighborhood world Jacobs envisioned (and experienced) was an historical artifact, a survival of the industrial city that had ceased to exist.

Thus while Jacobs was right to praise the associative possibilities of neighborhood, she failed to account for the effects of social and economic forces on the long-term survival of the neighborhoods she values. Murray Bookchin points out in *The Limits of the City* that the very "neighborhood world" Jacobs celebrated was in the process of being destroyed by "the same forces that truncate the inhabitant of the new town." Political and economic forces, he said, "are delivering the small shop over to the supermarket and the old tenement complex over to the aseptic high-rise superblock." This reminds us of the deep-seated economic and cultural forces that have made private *consumption* the focus of aesthetic concern. The contraction of the public sphere was linked to expansion of engineered private "experience"— and the new town was only one manifestation of this drift. In such a cultural and economic environment, the neighborhoods Jacobs admired "will continue to exist, but they will remain merely enclaves."[27]

Jacobs assumed that political organizing would of itself reverse this process; I think this assumption underlies her attempt to make irrelevant the entire planning discourse about the effect of spatial organization on social and cultural life. But again, turning to Bookchin's work, we have to remember that genuine decentralization must be seen in relation to the whole culture. And Bookchin invited us to question what kind of "communities" can survive when a city has become an "urban belt" that makes no provision for identification as place(s): "New York City, in fact, is no longer a city in the classical sense of the term and hardly rates as a municipality

[26]Marshall Berman, *All That's Solid Melts into Air* (Harmondsworth: Penguin, 1982), pp. 312–329.
[27]Murray Bookchin, *The Limits of the City* (New York: Harper & Row, 1974), p. 122.

even by nineteenth-century standards of urbanism. The 'city' is a sprawling urban belt that sucks up millions of people daily."[28] Under such geographic conditions a decentralized politics and economy is difficult to achieve. This reminds us of Mumford's earlier criticism of the *New York Regional Plan* and of the importance of creating decentralized *places* as well as fostering a democratic "society" and politics.

By not distinguishing between the "visionary" planning of Le Corbusier and that of Howard and Mumford, Jacobs failed to notice how different aesthetics imply different social visions. That both Mumford and Le Corbusier projected planned environments is less important than the sharp differences in their respective social visions. And these differences reflect the aesthetic contrast between a compact community and a high-rise-dominated city. The RPAA read from garden city and regionalist aesthetics a democratic social vision of "community" based on shared civic responsibilities. By contrast, the social order of Le Corbusier's *ville radieuse* was one of a rationalized bureaucratic regime that can efficiently manage the economy of mass production. For Le Corbusier, this necessitated a universal principle of rational design, a mathematical "measure" he called the "modulor" that is capable of creating "harmony" between the "human scale" and all "manufactured objects," including the built environment and consumable products.[29] Everything from the organic (the human body) to economic products, became part of a single field subject to a mathematical principle.

Le Corbusier did not separate this aesthetic principle of architecture from its larger context: the landscape/cityscape. The social principle inherent in this was implied in Le Corbusier's statement that "in the construction of objects [products] of domestic, industrial or commercial use, such as are manufactured, transported, and bought in all parts of the world, modern society lacks a common measure

[28]Murray Bookchin, *The Rise of Urbanization and the Decline of Citizenship* (San Francisco: Sierra Club Books, 1987), p. 246.

[29]Le Corbusier, *The Modulor*, trans. P. de Francia and A. Bostock (Cambridge: Harvard University Press, 1966; orig. pub. 1954), p. 60.

capable of ordering the dimensions of that which contains [i.e., the built environment] and that which is contained [i.e., that which is produced]."[30] To create such a common measure capable of subsuming all human activities in an industrial society is to create an environment that expresses the uniformity, efficiency, and rationality of mass production and consumption. Paul and Percival Goodman call this "a city of efficient consumption." Such a city is "a big community" (three million) in which "mass production provides the maximum quantity of goods"; it is carefully zoned into vast functional spheres "according to the acts of buying and using up." By breaking down the differences between modes of production and products, the logical consequence of Le Corbusier's aesthetics was to turn all experience into aspects of consumption. Not surprisingly, the Goodmans' "city of efficient consumption" organized space into zones of consumption that move outward from the center of the city, encompassing everything from consumable objects (the marketplace zone), to "truth" (the university and museum zone), to leisure, social intercourse and "nature" (the domestic zone).[31]

Whereas Le Corbusier's was an "imposed order" because it created the built environment according to the human image, Mumford's was an "organic" order because it connected aesthetics to the ecology of the natural region. The visual order of the garden city encompassed both the built environment as a "community" and the built environment in its relation to the ecology of the region. The aesthetics of the regional garden city was intimate and recovers a relation to the natural *region* because its original vision was one of decentralized production, intimate social relations, and the recovery of a "civic" sphere. It is "modern" in that it responded to "modernization": it used the opportunities created by neotechnics to create a new spatial order. Although this involved compromise, such as the zoning of industry as a response to large manufacturing plants, the social order remained civic.

Failing to recognize these differences, Jacobs's error has been repeated in recent academic studies. For example, in *Dreaming the Ra-*

[30]Ibid., 20–21.
[31]Paul and Percival Goodman, *Communitas* (New York: Vintage, 1960; orig. pub. 1948), pp. 125, 128.

tional City, Christine Boyer argued that urban planning as a profession and an intellectual tradition has been overwhelmingly "rational" or "disciplinary"; it dreamed a "perfect" order that it imposed by controlling "knowledge." And it is the exclusive access to knowledge that makes possible planning as social control.[32] Boyer's critique of power as knowledge is important in that it undermines the assumption of the authority of "experts" and shows why many modernisms fail because they overlook the complexities of social and cultural life. But there is also a serious limitation and danger in this kind of critique. It is evidenced in the way we seem to take the images created by the modernist architect-planners: sometimes too seriously and sometimes not seriously enough.

The critique of power as knowledge does not take the aesthetic images of the visionary planners seriously enough to see the important differences. Consequently, we may fail to appreciate the relation of "vision" to historical process: for all its limitations, the garden city as a planned visual order presented an opportunity to shape something different—and I think better—than what we have now.

The divergence between "rational humanist" and "organic" visual orders represented an important cultural divide. For Mumford, it provided a way to talk about our relation to the region and to other living beings. Garden city planning provided a way to frame a discourse about "nature." This may be its most important legacy. There is a politics to the "production" of aesthetic images, but there is also an important moral and visionary dimension that is necessary to shaping a successful polity.

At the same time these (modernist) images of planned cities are taken too seriously because they did not in themselves create the "anti-city." Quite the contrary. It is interesting to note the way in which conflicting visions *of transformation* have been appropriated to create the *existing* urban "order." For both the *ville radieuse* and the garden city have been plundered, the images of gleaming high rises amidst parklands and human-scaled towns surrounded by open country appropriated to further the disorderly spread of megalopolis. In the former we get "urban renewal" projects, "a col-

[32]M. Christine Boyer, *Dreaming the Rational City: The Myth of American City Planning* (Cambridge: MIT Press, 1983).

lection of high-rise slabs and towers linked by multi-laned express-ways"; and in the latter, "the pursuit of nature denatures the coun-tryside and mechanically scatters fragments of the city over the whole landscape" (*UP*, p. 142). Planners' images have become "cul-tural capital" to be spent by the "private" promoters of endless resi-dential developments outside the city and the "public" promoters of "urban renewal." The architect who creates the cultural capital is powerless to determine how it will be spent. So much for the vi-sionary-planner's authority. What one sees driving through America today is neither the efficient city nor the garden city. It is a land-scape given over to the task of accumulating power by means of destroying the potential of the ecological region—both in its "civic" and "natural" dimensions. The modernist images are not the source of this system. Yet many critics seem to find solace in blaming planners and visionaries for the forces that are destroying our cit-ies and our planet.[33]

[33]This reminds me of the pronounced tendency on the part of what is now called the "populist Right" to blame the profound social and eco-nomic woes of contemporary America on federal "bureaucrats" as if these bureaucracies were fully independent of corporate bureaucracies. Note also the way Jacobs's argument has been assumed by liberal pluralists who end up justifying the status quo. For example, in "Planning for People Not Buildings," *Environment and Planning* 1 (1969): 33–46; reprinted in M. Stewart, ed., *The City: Problems of Planning* (London: Penguin, 1972), pp. 363–384, Herbert Gans blames planners both for wanting "to put as many people as possible in single-family housing" (p. 369) and at the same time imposing an ideal of "urbanity" on people, an ideal that is resisted since most Americans, apparently completely free to choose their own destiny and independent of other cultural and commercial influences, "do not en-joy walking . . . do not like the city's congestion, and . . . do not care much for the heterogeneity of the population they find in the city" (p. 376). Rather than seeing that the planning debate might generate the prospect of find-ing a different way of life—whether that be defined in urban, regionalist, or civic terms—Gans advances a false populism that is actually conde-scending: people of the working and lower middle classes "are united in their opposition to the urbane upper middle class style of involvement in 'culture' and civic activity" (p. 377). Incredibly, Gans doesn't see these styles as constructions of culture and class but as reflections of popular will.

Nonetheless, Mumford recognized the importance of Jacobs's critique of planning. In a series of articles for the *Architectural Record*, Mumford acknowledged and criticizes her work. It is "very refreshing," especially necessary for planners whose "minds unduly fascinated by computers carefully confine themselves to asking only the kinds of questions that computers can answer."[34] Like Mumford, Jacobs challenges the scientific-rational planning paradigm that has dominated the profession since the time of Thomas Adams's *New York Regional Plan*. And Mumford concurred with her position that what is valuable about the city as a form comes through the process of accretion:

> An organic image of the city requires for its actual fulfillment a dimension that no single generation can ever supply: it requires time, not merely an individual lifetime, but many collective lifetimes. . . . The most valuable function of the city [is] as an organ of social memory; namely, its linking up the generations, its bringing into the present both the usable past and the desirable future.
>
> No organic urban design for any larger urban area accordingly can be completed once and for all, like a Baroque city established by royal fiat. (*UP*, pp. 161–162)

Mumford admitted that a valid urban design "must be one that allows for its historic and social complexity, and for its continued renewal and reintegration in time." The garden city suffered, then, not because it was planned but because its planning relied too much on a "single instantaneous image" (*UP*, p. 165). In this way Mumford confirmed Jacobs's judgment that a city cannot be a work of art that stands unaltered against time. He more or less conceded that the era of large-scale suburban development was receding. And he affirmed her disapproval of low densities and the overtly residential, as opposed to mixed-use, character of existing new towns. Now planning must address existing patterns of settlement, which is to say they must respond to historical precedents. In these respects

[34]The articles were reprinted in Mumford, *The Urban Prospect*, p. 186; hereafter referred to as *UP*.

Mumford acknowledged Jacobs's criticism of the garden city.

All this Mumford was ready to grant, but he also insisted on defending the garden city tradition—and *the validity of physical planning*—from what he characterized as "libelous caricature and ignorant abuses" (*UP*, p. 158). His central point was that a complete community does involve aesthetic questions, specifically the design of place. Without concern for the form and design of cities, there is little prospect for cultivated urban life. Mumford explained that we must distinguish between the "contents" and the "container" of the city. The "container" designates the form or structure; the "contents" include the people, their way of life, social organization, human memory, and cultural dispositions.

Mumford continued to stress the need to *design* viable urban forms that can channel human activities in a way compatible with the rediscovery of a functional and creative relation between culture and the natural world. For Mumford, the "container" must stand in creative relation to its "contents"—the human society that inhabits it—in a way that permits "organic complexity" (*UP*, p. 165). The most important lesson to be drawn from the garden city pertains to its vision of "place." Mumford stood up to Jacobs's ridicule of the garden city regional planning tradition because he understood that the larger issue is the importance of an aesthetic-moral perspective that informs the creation of real communities. When made part of a planner's vision, a sense of "place" fosters interaction between residence and work and between the built environment and the natural region.

Consequently, it is not surprising that during the 1960s Mumford continued to examine the prospects for regional land use planning. His focus shifted, however, from national to local and regional efforts. The federal effort to extend automobile suburbanization was just reaching its peak,[35] and the era of important environmental leg-

[35] According to Edward Soja, *Postmodern Geographies: The Reassertion of Space in Critical Social Theory* (London: Verso, 1989), the restructuring of the city after the Great Depression was a process in which "capital and the state worked effectively to replan the city as a consumption machine, transforming luxuries

islation had yet to unfold. Mumford feared that any solution coming from Washington would only further sprawl. "I don't think," he added, "that in a country as big as the United States, the financing of the New Towns by the Federal Government will prove any less of an obstacle to sound regional development than the subsidising of the National Highway Programme has been: on the contrary, it will only add to the existing sprawl and untidiness until a regional framework, embracing a whole state or a group of states, is put together by the states themselves."[36] The immediate solution, as demonstrated in *The City in History*, was to approach current planning initiatives (especially on the local and regional level) as practical steps in the right direction.

Mumford's continued interest in regional planning was evident in his continuing exchange of letters with Frederic Osborn and in the many practical planning proposals in *The City in History* and in his articles for the *Architectural Record*. One example is his assessment of planning efforts in Princeton, New Jersey. After visiting the town in 1964 on the occasion of a lecture, he shared his observations with Osborn. Mumford remarked that Princeton came closer to the New Town principle than contemporary wholly planned new towns. One difference was that Princeton had a well-established town center, including the university, which, Mumford noted, had been improved by restricting automobiles and locating parking facilities off the main street. The result was that Princeton had become a "genuine little city" of 11,000; a little city on its way to becoming a "balanced community" complete with a small urban core, an outlying factory district, and a de facto greenbelt consisting of outlying estates and university-owned lands.

In commenting on the lecture, Mumford told Osborn that "the audience was very responsive—all except the architecture and city

---

into necessities, as massive suburbanization created expanded markets in consumer durables" (pp. 101–102). This "State-Managed Urban System" enabled extensive suburbanization but also required "selective abandonment of the inner urban core" and a very expensive system of state subsidies to those who remained in the urban cores (pp. 180–181). The onset of another economic crisis in the early 1970s obviated this state-managed system.

[36]Mumford to Osborn, 10 April 1966, in *Letters of Mumford and Osborn*, pp. 401–402.

planning students, who for the most part are trained to connive at the irrationalities of our age." Yet even while seeking to elicit popular support for local and regional planning, Mumford also had to acknowledge its limitations. The audience, like the town itself, was made up of "people of means," many of them New York commuters. It would be easy to build support for greenbelt preservation among such a constituency. To mention that the town had failed to make "public provision" for workers' housing was another matter.[37]

Mumford's comments are important in two respects. First, he understood the limitations of local and even regional initiatives. Considered in isolation, such approaches will probably fail to address the fundamental social questions of class segregation and uneven access to necessary services. Second, while Mumford failed to criticize the strict functional separation of industry, residence, and commerce in the Princeton area (he had yet to see this as a limitation of New Town principles)[38], he did anticipate what has become today an important agenda in state and local planning: suburban recentralization and regional land use planning for the creation of open space.

Almost thirty years later many of Mumford's ideas are seen in new planning initiatives at the state and local level. For example, the State of New Jersey's *Development and Redevelopment Plan* favors centering new growth in suburban towns, creating town centers for other developing areas and preserving significant natural and working landscapes.[39] The new emphasis on regional planning is a response to the growing recognition of what Mumford, MacKaye, and the RPAA had prophesied in the 1920s: the spread of urban functions outward, a continuing process of decentralization, will itself become one of the most serious environmental threats we face. And to some extent, the new regional planning begins to recapture part of the RPAA legacy in another way. As Frank Popper points out, much of the planning that continues to dominate the profession is essentially

[37]See Mumford to Osborn, 13 November 1964; in *ibid.*, pp. 375–376.

[38]For an excellent criticism of the shape of the Princeton technoburb as a "corridor," see Whyte, *City*, pp. 299–306.

[39]New Jersey State Planning Commission, *Communities of Place: The State Development and Redevelopment Plan* (Trenton: State of New Jersey, 1989).

"regulatory." It makes use of the regulatory tools—zoning, master plans, environmental impact statements—of government at all levels. Despite political shifts that emphasize so-called free market approaches, this regulatory environment will by necessity continue to make itself felt.[40] The New Jersey state plan and similar approaches are innovative in the sense that they call for a better coordination between local and state planning—and in the case of the New Jersey plan actually create a mechanism to accomplish this. This is important because such plans are necessary attempts to establish long-term policies in place of piecemeal planning.

But real innovation goes beyond *regulation* (even systematic regulation) of urban decentralization, to the *prevention* of sprawl by shaping a viable recentralized economy in the emerging technoburbia.[41] (This is not to say, however, that to go beyond regulation is to transcend it, for the regulatory climate can help to stimulate the search for preventative approaches.) This preventative, or proactive, strategy recalls the original regionalist economic and cultural approach of the RPAA, which saw the necessity and desirability of concentrating new economic activity in decentralized centers. Mumford's comments about Princeton indicate that he had found a way to make the RPAA's legacy relevant to a country that has already experienced massive decentralization: the necessary movement is suburban *recentralization*. And Mumford realized that to be successful recentralization must go beyond professional *regulatory* planning—no matter how systematic—to a prevention strategy. Thus he insisted on the importance of a multimodal transportation system. The automobile, he pointed out in *The Urban Prospect*, creates only an "illusion of freedom"; "its utility decreases in direct proportion to its mass use, and in taking over the burden of public and private transportation, both passengers and freight, the motorcar has, with the aid of extravagant public subsidies, under the pretext of 'national defense,' wrecked the balanced transportation system that existed a genera-

[40]Frank J. Popper, "Understanding American Land Use Regulation Since 1970," *Journal of the American Planning Association* (Summer 1988): 291–301.
[41]I'm in debt to Barry Commoner, *Making Peace with the Planet* (New York: Pantheon, 1990), who applies the "control-prevention" approaches to pollution.

tion ago."[42] By re-creating a balanced transportation system, we could help generate the economies essential to recentralization. K. H. Schaeffer and Elliot Sclar have picked up this theme, arguing that the development of a multimodal regional transportation system would not only significantly advance land use planning but make employment and housing in technoburbia more accessible to urban populations otherwise tethered to inner-city neighborhoods, thereby reinvigorating the entire metropolitan region.[43]

Another proactive approach calls for the development of "walkable neighborhoods." Peter Calthorpe calls his new suburban planning technique "the residential pocket": a street-centered approach to create a mixed-use environment accessible to multimodal transportation.[44]

Policies and planning practices that aim at land use reform are important. The continually expanding "technoburb" is a point of weakness in the structure of the U.S. economy and polity. Dependent on supplies of cheap oil and usable land, promising a way of life too expensive for increasing numbers of Americans, sprawling and inefficient urban regions are the legacy of an American dream gone awry. They are the manifestation of an economic and environmental strategy that simply makes no sense in a world of shortages (scarcities not only of resources but of jobs). Strategies for recentralization and the development of multimodal transportation systems are necessary first steps in redesigning not only our built environment but our regional economies. "Place" is no aesthetic nicety: Mumford's realization that aesthetics, economy and polity must be rethought in relation to specific regions remains an important insight, and a fundamental point of departure.

[42]Mumford, "California and the Human Horizon," address before the Institute of Planning for the North Central Valley, at Davis, Calif., 12 January 1962; reprinted in UP, p. 11.

[43]K. H. Schaeffer and Elliott Sclar, Access for All: Transportation and Urban Growth (Harmondsworth: Penguin, 1975).

[44]See Peter Calthorpe, The Next American Metropolis: Ecology, Community and the American Dream (Princeton: Princeton Architectural Press, 1993). Calthorpe is an innovative California planner and an important advocate of a new kind of community planning often called the "new urbanism."

Conclusion:
The Relevance
of Ecological Regionalism

In the cultural and political milieu of our era, a period marked by disillusionment, the best work in city planning is associated with one aspect of postmodern "contextualism": the recovery of "urban universals."[1] In the American context this universal is preeminently the street.

William Whyte's, *City*, like the work of Jane Jacobs, seizes upon a great conservative truth, "the primacy of the street." As an ancient and very practical matter, cities need lively streets for a pedestrian life: "it is the river of life of the city, the place where we come together, the pathway to the center. It is the primary place."[2] In the older patterns of urban life (even in the industrial city), the pedestrian street retains its ancient social function of creating an urban

[1]See Charles Jencks, *Modern Movements in Architecture*, 2d ed. (Harmondsworth: Penguin, 1985), p. 374.

[2]William Whyte, *City: Rediscovering the Center* (New York: Doubleday, 1988), p. 7.

215

way of life by virtue of the constant human interaction made possible by its congestion and happenstance.

The need to restore the street is a powerful argument. I agree with Whyte that it is "sad to see how many cities have . . . emptiness at their core."[3] But in reviving the traditional urban contexts, must we forget the larger ecological/natural context? Beyond "urban humanism," beyond even the practice of *civitas*, the civic life of the Greek polis, there is the older linkage of eros, the interconnection of all life within the ecological region. The lived experience of place is a matter of contemplation, love. This is why, as the Goodmans remind us, whole places should include sites for contemplation as well as those dedicated to productive and social activity.[4]

The strong argument of Jacobs's and Whyte's work is undeniable. It can be seen in Paul Goldberger's introduction to the 1990 reprint edition of *Communitas*. Goldberger claims that the Goodmans imagine a city that is "truly *urban* . . . noble and civic and monumental and public."[5] Yet he understates another important aspect of the Goodmans' work, one that they share with Mumford: that the city must be reimagined in relation to the larger natural region that contains it.

The movement for "ecological restoration" is making an important contribution to the recovery of the natural region. Restoration means re-creating presettlement landscapes by using a very sophisticated ecological science to identify, classify, and reestablish "wildness." The wild usually conjures up images of vast "untouched" and remote places, but the *idea* of the wild rests on the reality of the ecosystem, a naturally occurring and self-sustaining balance of

[3]Ibid., p. 6.

[4]In *Communitas* (New York: Columbia University Press, 1990), pp. 153–186, Paul and Percival Goodman propose planning a city as the fullest expression of an "integrated community" around squares, some meant for productive activities, others for contemplation. For a suggestive handbook of "urban village" design and planning, see also Chris Alexander et al., *A Pattern Language: Towns, Buildings, Construction* (New York: Oxford University Press, 1977).

[5]Goodmans, *Communitas* p. x.

plant and animal life. Restoration means two paths to the creation of ecological systems within urban places: the "naturalization" of existing parkland and the reconstruction of "waste" land.

For the existence of wild patches within the city is more than a scientific or technical question; it is founded upon our realization that the wild once encompassed the land upon which our cities have been built. Bringing nature into the city, in both its wild and cultivated forms, is as much an act of mind and spirit as it is of physical planning. The creation of patches of wild places within a city summons our attention to the life-sustaining regional ecosystem that lies beneath the existing urban infrastructure. Narrowly defined, restoration is a question of landscaping, of converting existing parks, with their manicured look and high inputs of energy (high-maintenance horticultural landscapes) into natural areas that reflect the succession of plant life in the locality and that can support local fauna. The recognition that natural areas may exist within the city and that it is vitally important for us to re-create or sustain such areas implies that the old dualisms between human and natural, city and countryside, can be replaced with the recognition that even the most manmade of environments, the metropolis itself, is sustained by natural systems.

Restoring the land (specifically, taking back the wastelands that industrialization created) is a process of *re-creating* parks and open spaces on an ecological basis. It means restoring ecological health by making the park a "habitat," linking all open lands in regional systems by greenways and flyways, creating green passages into and through metropolis that will restore (wild)life. This may be extended to regionalism and regional planning in the sense that MacKaye used the terms: regionalism as an "exploration" of the indigenous environment, regional planning as redesign of the built environment to put it in relation to regional systems.

William Jordan explains that restoration is the basis of an "environmentalism that is ecological in tone and character"; restoration is focused on "relationships with and between creatures and the landscape" and sees landscape not as environment but as habitat. The difference is one of emphasis: environment implies an externalized world of "forces" to which creatures are subject; habitat implies the relation of all creatures to place.[6] Jordan calls restoration a "deed" rather than a "product"; it reflects a "primary concern . . . about relationships." Restoration is, Jordan reminds us, a "liturgy," a "ritual" that symbolically restores our citizenship in the community of all living things.[7]

Jordan's discussion of restoration helps put regionalism into ecological perspective recalling the most important element of Mumford's vision: that an authentic regionalism begins in the *experience of place* that generates and must find appropriate cultural or symbolic expression.

▣

Bringing culture into a reciprocal relation to nature was Mumford's primary concern. The garden city promised to reestablish this environmental balance, not only to preserve natural landscapes but to redesign the built environment. Regional planning of the sort that mattered meant shaping the whole metropolis and surrounding new garden cities with open countryside. Garden city regional planning addressed the importance of overcoming the fragmented landscapes—urban, rural, and wild—created by the revolutions in agriculture and industry.

It may be said that as the chief instrument of RPAA planning practice, the garden city approach to regional design was oversimplified.[8] Looking back at RPAA design equipped with current ecological thinking, we can easily see the RPAA's limitations. But to criticize garden city regional planning for its lack of ecological sophistication is to miss recognizing Mumford and MacKaye as innovators of the *notion*

---

[6]William Jordan, "Making an Urban Wilderness: Reflections on the First Fifty Years of the University of Wisconsin Arboretum," in Gordon, *Green Cities*, pp. 67–78; see also Mark Luccarelli, "Review of *Green Cities*," *Our Generation* 22 (Fall 1990–Spring 1991): 132–136.

[7]Jordan, "Making an Urban Wilderness," p. 77.

[8]The best statement of this idea is found in William Whyte, *The Last Landscape* (Garden City, N.Y.: Doubleday, 1968), pp. 135–162; see Chapter 6, footnote 3 for further discussion.

of ecological planning. Furthermore, the use of the garden city idea by the RPAA was a response to the exigencies of history. The RPAA anticipated the coming suburban exodus, and the garden city was a pragmatic attempt to address this coming crisis and opportunity..

The regionalist imperative—the preservation of countryside and wild places—is best realized today not through the garden city idea of creating greenbelts to keep the places of megalopolis geographically distinct (largely a lost battle) but through a process of reconfiguring the existing built environment. The tools for doing this include setting aside ecologically significant lands, preserving and redeveloping regional agriculture, and creating "greenways" and trailways in and around cities.

But the larger task is to create a built environment—a cityscape, that reflects and in some significant way incorporates the wild and rural landscapes. We must create "green cities," restoring significant (and symbolic) places for biological diversity. By restoring wastelands and ceasing to poison the environment, we can make room for different species; many species of plants and animals have shown great tolerance for human activity. We can restore an ecological landscape within and through the urban areas. This will make possible the connection of cities, both physically and imaginatively, to the region and the biosphere.[9]

The ecological regionalism Mumford pointed us toward is a *practice* that permits us to imagine and create a new relation between production and consumption, between function and aesthetics. This reminds us that for Mumford regionalism was an "occasion of experience," and a *vision* of moral and aesthetic renewal as well as strategy for reforming (and re-creating) political economy. For Mumford, political economy and technology were fundamental, because he saw his work as a response to the historical possibilities created by "neotechnics." Later he became convinced of the need to make the revitalized public sphere essential to reviving regional economies. Regionalism demands the recovery of the polis, the reinvention of the economy.

[9]An interesting and useful guide on this kind of planning is Peter Berg, Beryl Magilavy, and Seth Zuckerman, *A Green City Program for San Francisco Area Cities and Towns* (San Francisco: Planet Drum Books, 1989).

The story of the RPAA's ideas as translated into New Deal policies is instructive here. The RPAA had failed in two senses: first it did not generate public interest in regional planning necessary to marshall support from below for its program. Second, its conception of a regional political economy simply lost out to competing theories of economic planning. This was best illustrated in the debate over the direction of the TVA (see pp. 156–161).

The revival of interest in regional land use planning indicates the continuation of Mumford's legacy: the importance of reorienting "place" as a means of social and environmental reform. In *The Next American Metropolis*, Peter Calthorpe maintains that the "form and identity of the metropolis must integrate historic context, unique ecologies, and a comprehensive regional structure."[10] To realize these goals, Calthorpe argues that we must establish "long-term regional policies." These are based on two important principles: (1) "land use policy reform," necessary to define established boundaries for all metropolitan areas, and (2) public-private partnerships to concentrate new growth in suburban hubs. These recentralized areas would be planned as "communities . . . designed to reestablish and reinforce the public domain." By expanding the mass transit network, "Transit-Oriented Development" could simultaneously reduce the over-reliance on the automobile and create suburban centers, *urban* places that are "human-scaled and . . . diverse in use and population."[11]

Calthorpe's version of the "new urbanism" is the best tradition of architectural reform—the very tradition Mumford defended against Jane Jacobs. Calthorpe's is a sorely needed voice in favor of the power of "place" to profoundly effect human experience. It is remarkable, too, how similar his goals for regional planning and suburban recentralization are to those conveyed by the RPAA sixty years ago. He has added an emphasis on urbanity (lively streets, mixed-use neighbor-

[10]Peter Calthorpe, *The Next American Metropolis* (Princeton: Princeton Architectural Press, 1993), p. 15.
[11]Ibid., pp. 15, 35.

hoods) that was lacking in the RPAA but was largely corrected in Mumford's work in the 1960s. But more importantly, like Mumford and the RPAA, Calthorpe brings together ecology and community in arguing for a complete geographic restructuring of the urban regions. Reminiscent of Stein's approach to reform in New York State in the 1920s, Calthorpe hopes to put together a coalition of private and public interests: environmentalists, enlightened developers, and inner-city advocates.[12]

Calthorpe's successful completion of several development projects combined with the advances in regional land use planning in some metropolitan areas (such as Portland, Oregon, and Montgomery County, Maryland) may be due to the increased visibility of the world-wide environmental crisis and the widening agreement that something must be done to address the contradictions of the American metropolis. But the hope for a revival of regional planning as a way to achieve what Mumford called "the regional framework of civilization" must be tempered by an awareness of the forces arrayed against it: the intransigence of the private sector, the continuing political attack against the agency of government in particular and the public sphere in general (accompanied by the revival of laissez-faire ideology), and most importantly the accelerating cultural detachment of a wide segment of the American public that comes from the physical isolation in technoburbia, augmented by the new importance of "electronic space that requires no bodily human presence or physical movement."[13]

In light of these factors and considering the experience of the RPAA, it may well be that if we are to make the experience of "place" matter, more fundamental changes than those contemplated in regional planning and land use reform might be necessary. Mumford's vision of ecological regionalism points to two further elaborations of a progressive politics of place: the idea of economic regionalism, and the revival of democracy.

One way to approach to this alternative, a political economy that

[12]Ibid., p. 36.
[13]Langdon Winner, "Silicon Valley Mystery House," in Michael Sorkin, ed., *Variations on a Theme Park: The New American City and the End of Public Space* (New York: Hill and Wang, 1992), p. 55.

takes place into account, is to begin to work to create new institutions at the local regional level. Gar Alperovitz suggests that we create community-owned industries that build an alternative to the "nonmarket" of international capitalism. Such institutions would provide the constituencies needed for meaningful urban and regional planning. The application of planning principles could create significant employment: "in the future, a public decision to build mass-transit systems, solar collectors, and recycling equipment as part of an overall plan for ecological balance could produce contracts and jobs which, in turn, could be targeted for community-stabilizing purposes."[14] Thus planning could be part of a democratic debate. Some U.S. cities tried such experiments in the 1970s, as Pierre Clavel describes in his important book *The Progressive City*. Clavel reveals the possibilities of progressive planning linked to decentralized economics and politics.[15] This book recalls Mumford's early work, specifically, his praise for the idea of communal "corporations"; and it coincides with some recent work, such as that of Christopher and Hazel Gunn in response to the question "what, in a global economy, could a community do for itself?"[16] and of David Morris on creating neighborhood, city, and regional economies and opportunities for locally owned firms to engage in international trade.[17] These studies suggest that new institutions—and new economies—may come to challenge or reinvigorate political structures.

The economic rationale for political change must be understood in relation to the destructive shift in the world's political economy. The shift has created two salient (and related) issues: the deindustrialization of whole regional economies (beginning with the

[14]Gar Alperovitz, "Building a Living Democracy: A Whole New Way of Thinking About Politics and Economics," in *Sojourners* (July 1990): 11–23, quotation on 19.

[15]Pierre Clavel, *The Progressive City: Planning and Participation* (New Brunswick, N.J.: Rutgers University Press, 1986).

[16]Christopher and Hazel Gunn, *Reclaiming Capital: Democratic Initiatives and Community Development* (Ithaca: Cornell University Press, 1991), vii.

[17]David Morris, "The Ecological City as the Self-Reliant City," in David Gordon, ed., *Green Cities: Ecologically Sound Approaches to Urban Space* (Montreal: Black Rose Books, 1990).

old industrial regions dependent on the consumer economy and continuing with newer regional economies based on defense industries) and environmental degradation, especially through resource depletion (which affects all regions but is particularly devastating in peripheral areas dependent on farming and extractive industries). Thus two fertile territories in which to begin exploration are in the declining urban/industrial regions and in the exhausted peripheral regions, both of which are so desperately in need of renewal. As Benton MacKaye reminded us, exploration entails the development of a region's indigenous resources. And reinvention begins with reassessment: what are the regional resources and how might a new "green" economy be shaped around them? This opens up the possibility of developing "green industrialism," a fundamental shift in the *rationale* of production, which requires that local and regional efforts inform a larger national political effort.[18]

Properly understood, green industrialism contests the psychological and ideological terrain upon which "economy" is predicated. This enables redefinition of the boundaries of economy to encompass ecological sustainability and regional design.

Participatory democracy could emerge alongside the development of regional economies. This might provide a direction for the question Dewey wrestled with in *The Public and Its Problems*: how to create democratic "publics" capable not only of enhancing civic life but of confronting the issues necessary to defining, and redefining, the public interest. As Mumford showed in his description of early New England, the region, conceived in its unit as both an aesthetic *and* economic unit, has the capacity to symbolically mediate between the social world and the natural ecology. The polis that was once the origin of democratic citizenship could well become the birthplace of ecological citizenship. Mumford's understanding that democracy meant tolerance for diversity in the context of defining a common or "public" interest became his most important metaphor for mediating the concerns of the social and natural worlds.

[18]On the possibilities of linking "green industrialism" with economic conversion of defense industries, see Jonathan Feldman, "Broadening the Peace Dividend," *Society* 30 (May/June 1993): 32–40.

Index